Moving the Construction Safety Climate Forward in Developing Countries

The construction industry in developing economies is responsible for creating deliverables such as infrastructure and housing while providing a means of livelihood to an ever-increasing number of management and frontline workers. However, in many parts of the world, injuries and fatalities have continued to damage the industry's image.

This book intends to meet the needs of many construction managers who, though technically informed, struggle with managing frontline workers, especially regarding motivating positive safety outcomes. It discusses the challenges experienced in the industry and how site management may navigate them to improve safety performance in the workplace. By documenting the experiences of site management in developing countries, this book intends to contribute to the education of professionals on evolving better safety environments on construction sites. It considers the safety climate in a high-risk work environment, administrative procedures and the implementation mechanisms. The book also documents findings from existing literature about developing countries in contrast to what is obtainable in developed countries. Each chapter features context-specific explanations from empirical research conducted in developing countries. Key safety climate issues are contextualised, considering the challenges faced in developing countries, alongside current trends that will help chart future directions that will promote continuous improvement of safety outcomes of construction projects.

This book is essential reading for construction managers, researchers and academics in the field of safety management, infrastructure delivery and project management.

Tchad Sharon Jatau lectures in the Department of Quantity Surveying at Kaduna State University, Nigeria.

Fidelis Emuze is Professor and Department Head in the Department of Built Environment at the Central University of Technology, Free State, South Africa.

John Smallwood is Professor of Construction Management at Nelson Mandela University, South Africa.

Routledge Research Collections for Construction in Developing Countries

Series Editors: Clinton Aigbavboa, Wellington Thwala, Chimay Anumba, David Edwards

For more information about this series, please visit: www.routledge.com/Routledge-Research-Collections-for-Construction-in-Developing-Countries/book-series/RRCDC

Moving the Construction Safety Climate Forward in Developing Countries

Tchad Sharon Jatau, Fidelis Emuze and John Smallwood

Routledge
Taylor & Francis Group

LONDON AND NEW YORK

First published 2023
by Routledge
4 Park Square, Milton Park, Abingdon, Oxon OX14 4RN

and by Routledge
605 Third Avenue, New York, NY 10158

Routledge is an imprint of the Taylor & Francis Group, an informa business

British Library Cataloguing-in-Publication Data
A catalogue record for this book is available from the British Library

Library of Congress Cataloging-in-Publication Data
Names: Jatau, Tchad Sharon, author. | Emuze, Fidelis, author. |
    Smallwood, John Julian, author.
Title: Moving the construction safety climate forward in developing
    countries / Tchad Sharon Jatau, Fidelis Emuze, John Smallwood.
Description: New York : Routledge, 2023. | Series: Routledge research
    collections for construction in developing countries | Includes
    bibliographical references and index.
Identifiers: LCCN 2022056077 | ISBN 9781032419220 (hbk) |
    ISBN 9781032421964 (pbk) | ISBN 9781003361640 (ebk)
Subjects: LCSH: Building—Developing countries—Safety measures.
Classification: LCC TH443 .J38 2023 | DDC 624.028/9091724—dc23/
    eng/20230111
LC record available at https://lccn.loc.gov/2022056077

ISBN: 978-1-032-41922-0 (hbk)
ISBN: 978-1-032-42196-4 (pbk)
ISBN: 978-1-003-36164-0 (ebk)

DOI: 10.1201/9781003361640

Typeset in Times New Roman
by Apex CoVantage, LLC

Contents

Figures

Chapter 1

Chapter 5

Chapter 6

Tables

Preface

We wrote this book for construction scholars and practitioners in developing countries. The idea is to meet the needs of construction professionals who, though technically informed, struggle with managing frontline workers, especially regarding motivating positive safety outcomes. We identify with the issues faced in construction because, by education and professional practice, we are scholars and practitioners with several years of exposure to health, safety and well-being issues in the construction space. Starting with his master's and doctoral studies, Professor Smallwood, for example, has addressed health and safety (H&S) issues for several decades using both industry and academic platforms. Over the past years, Professor Emuze has taught and researched H&S issues at the operational and strategic levels. He designed university programmes on H&S in South Africa. The goal of this collective effort is to ensure that the worksite does not harm people in construction. For a site to be free from harm, the climate, which summarises the perceptions of operatives in a work environment, must be suitable. Though the safety climate has enjoyed a pivotal level of interest in developed countries, it is grossly underreported in developing countries. The book intends to close this gap.

The book showcases the experiences of site management in developing countries, with Nigeria and South Africa as case examples. It discusses the challenges experienced in construction and how site management may navigate them to improve safety performance in the workplace. The book presents findings from existing literature and empirical data from these two developing countries to foreground the significance of safety climate. Each chapter features context-specific explanations from the literature and the empirical research. The chapters unpack the challenges which developing countries face and make suggestions that will promote continuous improvement of safety outcomes of construction projects. The empirical information provided in each chapter emanates from a single study. The research methodology for the study is therefore summarised here.

About the Research Methodology

The paradigm for the research conducted in two developing countries (Nigeria and South Africa) is pragmatism. The world view enabled Dr Jatau to use varying approaches as directed by the research problem and questions for new knowledge incidental to her doctoral study. The pragmatic position allowed us to view the problem from multiple angles. The pragmatic position led to the use of mixed methods, which enabled the collection of both textual and statistical data from both countries. The sequential explanatory mixed-methods research design facilitated the collection of quantitative data before the qualitative data. The results of the quantitative data (survey) led to the compilation of the questions used to collect the qualitative data (interviews). The sample for the study was drawn from professional construction groups in Nigeria and South Africa. The South African participants were drawn from the South African Council for the Project and Construction Management Profession. In contrast, the Nigerian group was drawn from the Architect Registration Council of Nigeria, the Council of Registered Builders of Nigeria, the Quantity Surveyors Registration Board of Nigeria and the Nigerian Institute of Civil Engineers. The purposively selected participants have minimum H&S literacy levels required to respond to the survey and interview questions. In all, 251 survey responses were analysed (quantitative strand), while 25 textual interview data were analysed (qualitative strand). The job profile of the sample shows that the survey respondents and the interviewees were architects, builders, construction managers, quantity surveyors, civil engineers, H&S officers and project managers.

The principal data collection instrument for the study was the Nordic Safety Climate Questionnaire (NOSACQ-50), which is widely accepted by scholars. Evidence of its acceptance is the translation of the same instruments into several languages (Chinese, Czech, Danish, Dutch, Belgium, English, Estonia, Finnish, Italian, German, Hungarian, Icelandic, Indonesian, Lithuanian, Malay, Persian, Polish, Russian and Turkish) to measure safety climate in high-risk work environments. The instrument is suitable for the study as it has been previously used among construction workers. The need for modification to suit the study was minimal, thereby preserving the instrument's validity and ensuring that the appropriate data were collected. The scale finds usability across different nations, starting from the Nordic countries. This suggests that the scale accommodates the differences that arise from the uniqueness of cultures and that it measures what it claims. The scale in the instrument deployed to measure safety climate in the two developing countries consists of 50 items across seven H&S climate dimensions. These dimensions include management safety priority, commitment and competence; management safety empowerment; management safety justice; workers' safety commitment; workers' safety priority and risk non-acceptance and safety communication, learning and trust in co-workers' safety competence and trust in the efficacy of safety systems. Accordingly, chapters in this book follow the dimensions of the NOSACQ-50 scale, as elaborated later.

About the Chapters in the Book

An overview of safety climate opens the book in Chapter 1, which reinforces the well-reported notion that hazard and risk exposure is a challenge to frontline construction operatives. The types of exposure and the likelihood of harm to people increase with labour-intensive operations, mostly found in the Global South, where most developing nations are situated. Chapter 1 of the book indicates how construction workers are faced with the risk of accidents when the workplace culture and climate are conducive to unsafe acts and conditions. The chapter thus describes the safety climate and emphasises challenges faced in its positive promotion on a typical construction site. The chapter further links safety climate to concepts such as safety performance and compliance, counterproductive work behaviour and organisational citizenship behaviour (OCB).

From the broad to the specific, Chapter 2 focuses on safety prioritisation in the workplace. The chapter reiterates the argument that management must engage in safety-related behaviours that show workers that unsafe behaviours are unacceptable. The core of the discourse in Chapter 2 is that safety priority is not an operational matter for management. Instead, it is a strategic issue that must permeate every level of an organisation. When frontline workers see top management 'walking the talk' in safety, they accept safety as a priority in the worksite and adopt the behaviours modelled before them. Chapter 2 also presents empirical findings from the two developing countries on management safety priority to underscore the idea that management acts influence safety-related behaviours of workers through OCB. The data in the chapter imply that management in construction should be aware of the significant role they play in promoting H&S among frontline workers. Positive roles foster an excellent safety climate and performance, while the opposite will happen with negative roles.

After priority, Chapter 3 of the book addresses empowerment. The chapter does not pretend to cover all that concerns empowerment. Rather, Chapter 3 foregrounds the notion that when management empowers its employees in H&S issues, a positive workplace climate is possible. Such an atmosphere improves the workers' ability to engage in safe acts and choices, which lays the foundation for better H&S practice. The chapter therefore reinforces safety empowerment among frontline construction workers. It argues that empowerment in construction could be achieved when management improves safety knowledge through training and related mechanisms. Safety empowerment is a win–win situation for both management and their workers. For instance, when frontline workers are empowered, management confidently delegates authority through subgroup leadership. This increases workers' perception of control over their work environment and enriches the decision-making process.

Safety justice is addressed in Chapter 4. Relying on the just culture discourse in safety science, the chapter contends that a just culture in construction would ensure that workers should be confident that measures taken by management to resolve or manage either human error or violations would be fair. The assumption is based on the premise that when management fails to demonstrate visible justice, frontline workers may show low levels of safety voice and incident reporting. The chapter highlights the benefits of safety justice in construction and suggests how management may explore it for the benefit of improved safety performance. The chapter supports the assumptions with empirical results from Nigeria and

South Africa. The study's outcome reveals that levels of safety justice in these developing countries are low and need to be improved.

Leadership, which is a concept that is gaining traction in organisational management, is the focus of Chapter 5. The chapter indicates that workers who are assigned leadership responsibilities in their workgroups will ensure that H&S is sustained for improved organisational performance. The chapter also argues that safety leadership must be provided so that supervisors can predict behavioural outcomes in an identified safety climate. In addition, Chapter 5 highlights three leadership styles (such as transactional and transformational leadership), which can be adopted by managers and supervisors to drive frontline workers towards improved safety compliance. Regardless of style, an ideal safety leadership ensures that frontline construction workers at group levels can champion the benefits of safety regulations compliance and foster mentorship among peers. Safety leadership at lower levels also enables workers to communicate to management when it fails to create an enabling environment for safety compliance to thrive.

An effort to foster a good safety climate will fail without the commitment of stakeholders. As such, Chapter 6 deals with safety commitment, which correlates with improved safety performance. The chapter implores management to adhere visibly to H&S as this communicates its importance to the workers, given that workers imitate those they admire. For example, the foreperson on construction site may be the only form of management with whom some frontline workers may get to interact; it is therefore practical that such individuals are committed to H&S. Safety commitment, however, goes both ways – it applies to both management and frontline workers. To improve safety commitment levels, management must invest in workers' safety knowledge and competence.

Chapter 7 focuses on effective communication in the workplace, which is the lifeline of performance. This chapter shows how safety communication influences safety learning and education. It argues that learning should extend beyond management to workers since safety education also occurs among co-workers. Management must explore approaches encouraging solid social cohesion among workers since it is necessary for safe climate evolution. Effective safety communication is an equal responsibility between management and workers and must be designed to allow a feedback mechanism (it must be bidirectional, not top-down). Safety communication improves skills and competence among all construction workers; by adopting non-traditional training methods such as teaching aids and non-verbal communication where necessary, safety education and compliance levels may be influenced in favour of improved safety performance. It is necessary that the management communicates with workers in a language that they understand. Chapter 7 underlines this reality in day-to-day construction operations.

The concluding chapter of the book is on trust. Chapter 8 reinforces the idea that the trust between management and workers contributes to a smooth relationship where everyone tries to fulfil their commitments. Where there is trust, there is less 'blame game' when H&S matters eventuate. The chapter thus highlights the role of establishing safety trust in the relationship between management and frontline workers in high-risk workplaces such as construction sites. The chapter concludes that the role of effective safety leadership, commitment and priority cannot be overemphasised in establishing safety trust in a construction workplace. Management must ensure that safety trust among workers is sustained so they can reap the benefits thereof.

In sum, the book covers priority, empowerment, justice, leadership, commitment, communication and trust within the context of what is required to foster an appropriate safety climate for everyone involved in construction in developing countries.

Acknowledgements

We are blessed with the expertise of associates in the construction health and safety field, who reviewed the chapters in the book. Their suggestions enhanced the book's quality. We are grateful to Andrea Jia (University of Melbourne, Melbourne, Australia), Chioma Okoro (University of Johannesburg, Johannesburg, South Africa), Ciaran McAleenan (Educational Consultant, recently retired from Ulster University, Belfast, UK), Innocent Musonda (University of Johannesburg, Johnnesburg, South Africa), Irewolede Ijaola (Yaba College of Technology, Lagos, Nigeria), Justice Williams (Appiah-Menka University of Skills Training and Entrepreneurial Development, Kumasi, Ghana), Kofi Agyekum (Kwame Nkrumah University of Science and Technology, Kumasi, Ghana), Philip McAleenan (Expert Ease International, Belfast, UK), Nicholas Tymvios (Bucknell University, Pennsylvania, USA), Nnedinma Umeokafor (University of Greenwich, Kent, UK), Pham Hai Chien (Ton Duc Thang University, Ho Chi Minh City, Vietnam), Rita Zhang (Royal Melbourne Institute of Technology, Melbourne, Australia), Sheyla Mara Baptista Serra (Federal University of Sao Carlos), Tariq Umar (University of the West of England, Bristol, UK) and Zakari Mustapha (Cape Coast Technical University, Cape Coast, Ghana). For the English language editing of each chapter, we appreciate the assistance of Renée van der Merwe. At Taylor and Francis in London, we thank Martha Luke and Ed Needle for their support.

Fidelis Emuze
Bloemfontein, South Africa
19 October 2022

Acknowledgements

1 Safety Climate in Construction

1.1 Introduction

The construction industry is responsible for providing infrastructure and jobs to an increasing percentage of poor job seekers. According to the United Nations, projections relating to world population increase indicate that by the year 2030, for every five people across the globe, three will be Africans in developing countries. The ripple effect of this projection will mean that the industry will be under immense pressure to deliver on infrastructure demands (Cheng et al., 2010). This implies that as demand in the construction industry increases, issues regarding the health and safety (H&S) of workers in developing nations must be taken more seriously.

The construction industry has been plagued by reports of accidents, which label it as one of the unsafe industries to work in; it is therefore expedient that all hands are on deck to improve safety performance on all levels. Globalisation and urbanisation have seen major cities requiring infrastructural development. Increasing construction requirements in developing nations means additional general workers are being placed on project sites with implications for H&S outcomes (Liang & Zhang, 2019). A developing region where this idea is most evident is sub-Saharan Africa (SSA). Thus, ensuring that the safety climate in SSA performs optimally can be seen as ensuring that this vulnerable category is catered for in a work environment that allows them to grow and provides commensurate job satisfaction and dignity. This section of the chapter provides an overview of construction H&S in SSA.

1.2 Construction Safety in SSA

Safety climate is generally described as workers' shared perception of the priority that management places on their H&S in the workplace. A safety climate report may be termed incomplete without an accompanying sister concept, namely safety culture. This is roughly described as the outcome of workers' H&S-related beliefs, attitudes, competencies and behaviours sustained over time (Suraya et al., 2022). Yorio et al. (2019) opined that an organisation's safety culture may be influenced by a broader prevailing national culture, given that management

DOI: 10.1201/9781003361640-1

safety values, beliefs, norms and behaviour reflect their culture. It can be said that the culture of a nation may determine management's safety culture approaches in construction. Countries in SSA have unique rich cultures and the people are majorly communal. A healthy safety climate and culture have been shown to aid high-risk work environments in improving their safety performance outcomes. Recently, the construction industry in Africa has proven to be valuable to the individual nations' economy. For example, it is an indicator which reveals a country's economic state, and it also provides a source of livelihood to a significant percentage of the populace.

It is pertinent to note that though the construction industry is vital to global economies in the provision of not only infrastructure and jobs, it is also reported to be one of the most dangerous industries in which to work (Fang et al., 2020). This implies that more accidents and injuries linked to construction will continue to occur if left unchecked. Fatality figures have continued to soar, leaving in their wake outcomes such as more injuries, disabilities and, in worse cases, death (Yang et al., 2012). The International Labour Organization indicated that worldwide annually about 340 million occupational accidents, with approximately 6,000 deaths daily, are attributed to workplaces. They added that the construction industry accounts for a high percentage of these accidents. Factors such as underreporting and the lack of a central database to monitor occurrences of construction accidents, among other factors, make it difficult to track accident statistics in the developing countries (Lestari et al., 2020). Yet in countries such as South Africa, where there is some statistical evidence of construction accident records, scholars still bemoan such data as a poor reflection of actual figures (Okonkwo, 2019). In Ghana and Nigeria, similar reports exist. What is pertinent to note is that construction accidents in SSA are grossly underreported (Alkanam & Afatsawu, 2022). This phenomenon may be attributed to poor levels of compliance and enforcement of H&S regulations on construction worksites, according to Umeokafor (2018). There may be evidence of self-regulating among contractors in countries such as Nigeria. Where there are no consequences, contravention of the law will abound.

Major investigations around H&S issues have emanated in a bid to ensure an 'accident-free' workplace. The literature reveals a plethora of accident causes in construction activities (Sanni-Anibire et al., 2020; Zwetsloot et al., 2017). For example, in the United Kingdom, construction workers are involved in 70 per cent of all reported accidents; issues regarding the workplace accounted for 49 per cent; problems with equipment (including PPE) amount to 56 per cent; suitability and material conditions were blamed for in 27 per cent of the accidents; while deficits in management accounted for 84 per cent of the accidents. Asanka and Ranasinghe (2015) identified human-related factors, such as negligence, as a dominant cause of accidents on construction sites. Winge et al. (2019), in another study, identified other factors responsible for accidents on construction sites to include workers' capabilities, poor communication, low levels of knowledge and skills, poor supervision, fatigue, site layout and space, housekeeping and workplace safety culture. This is worse in SSA countries where harm to workers occurs daily (Awwad et al., 2016).

Developing countries are faced with a myriad of construction-related H&S challenges, with evidence of poor interest and a lack of commitment to the subject from stakeholders such as researchers (Umeokafor & Isaac, 2016). In Kenya, for example, Olutende et al. (2021) bemoaned the fact that casual construction workers' rights are often trampled upon by contractors owing to the nature of their employment tenure. They added that the Trade Dispute Act of 1991 does not protect this category of workers. As a result, contractors get away with under-reporting of accidents and the creation of unsafe work environments which increases the likelihood of accident occurrence. Boadu et al. (2021) blamed the poor state of occupational health and safety (OHS) in Ghana for the non-enforcement of regulations. The picture is no different in countries such as Nigeria, which still struggles with the non-existence of a working OHS regulation tailored for the industry, among other factors such as bribery, corruption and over-inflation of contract cost (Umeokafor et al., 2018).

In South Africa, there is evidence of commendable research outcomes on how to improve construction H&S. However, this has not been matched by commensurate enforcement and compliance (Windapo & Oladapo, 2012; Awwad et al., 2016). Further, they added that the poor documentation of accidents and near misses have made it impossible to track the statistics. In Nigeria, evidence from the literature confirms that the prevailing H&S policy is embedded in the Factories Act, which originated from the United Kingdom, and does not consider the country's environmental peculiarities. As a result, contractors who engage in any form of H&S implementation self-regulate (Umeokafor, 2017). This may imply that the contractors' knowledge (often limited) of acceptable standards may determine their H&S regulation at work and this is generally unacceptable.

Chim et al. (2018) classified the causes of construction accidents as both natural and the faults of people. Accidents attributed to the faults of humans may include failure to enforce procedures, inadequate facilities, failure to recognise hazards, failure to motivate, poor design and selection, poor maintenance, inadequate instructions, inappropriate training methods, poor attitude and work procedures and inadequate planning and layout (Gibb et al., 2014; Hosseinian & Torghabeh, 2012; Chim et al., 2018). Construction workers are exposed to different types of accidents or injuries in their line of duty; as such, the safety climate and culture in place may influence their perception of the likelihood of occurrence of such accidents and their tendency to engage in unsafe H&S practice (Emuze et al., 2016). Though research has been conducted to identify accident causes, there is still much to be done in terms of understanding unsafe behaviour among construction workers (Mohammadi et al., 2018).

The safety climate of a work environment influences risk behaviour and the likelihood of injury in the workplace. Empirical studies have confirmed that accidents and near misses are significantly related to human error; for example, Guo et al. (2016) stated that workers are often blamed for forgetfulness, inattention, incompetence and lazy attitudes when accidents occur. They added that underlying factors which stem from the organisation are capable of modifying human behaviour in the long run. According to Fu et al. (2020), accidents attributable to humans may be a function of individual tendencies, as some individuals have more accident characteristics than others. For example, Fu et al. (2020) mentioned that factors such as different human personality types may contribute to workers' engaging in behaviour that increase their likelihood of being involved in an accident; individuals experiencing high stress levels may be involved in an accident and gender may also be a predictor of accidents with males having a higher chance of accident involvement than females. Emuze (2018) added similar factors as causes of human error to include poor judgement, failed techniques, inexperience, poor communication, poor monitoring and control and insufficient preoperative preparation, among others.

Seeing that the workers' behaviour plays a vital role in accident causation, it is important to understand the role of management as well. According to Umeokafor (2017), management has shown much disregard for the safety of workers, assigning a higher priority to other work-related objectives such as productivity, meeting of deadlines and profit maximisation. Okorie (2014) suggested strong management and leadership skills as a requirement in ensuring that construction workers perform optimally along H&S lines, given that the quality of top leadership directly influences construction workers' behaviour.

Dane and Brummel (2014) added that construction workers are equally concerned with meeting set productivity targets and when overwhelmed, they tend to make certain negligent decisions in a bid to accomplish targets set by their managers. Frontline workers who perceive that productivity is recognised or rewarded in the workplace over H&S goals may engage in behaviour they perceive meets their employers' expectation while ignoring their H&S. For example, when production pressure influences managers to make speed of completion appear as a stronger priority over workers' H&S, workers tend to adopt work methods they perceive as faster to impress their managers and avoid any negative consequences (Guo et al., 2016). This may result in grave consequences linked to the causes of accidents on the worksites.

1.2.1 Non-compliance of OHS Regulations

OSH is an all-encompassing concept that applies to work sectors such as mining, health, manufacturing, aviation, construction, chemical, nuclear energy and food production, among others. It covers workers' mental, emotional and physical well-being in the workplace (Amponsah-Tawiah & Mensah, 2016). In South Africa, compliance to OSH regulations is deemed compulsory as stipulated in the Occupational Health and Safety Amendment Act 85 of 1993, which generally provides for the H&S protection of persons in connection with the use of plant, machinery and hazards in the workplace. The Act further caters to the safety of other persons such as passers-by and residents who may be exposed to hazards arising from the activities of workers. Violations of this regulation are subjected to penalties stipulated in the decree.

The OSH Act of 1993 provides for the Construction Regulation of 2014 as a decree which regulates construction activity in South Africa and was published in the *South African Labour Guide*, 2016. The Construction Regulation decree states the responsibility of all parties involved in the construction activity, as stated in the Construction Regulation 7(1)(a) and 7(1)(b) (South African Government, 2014: 15). First among the duties of the principal contractor is to:

- *provide and demonstrate to the client a suitable, sufficiently documented, and coherent site-specific H&S plan, based on the client's documented H&S specifications contemplated in regulation 5(1)(b). The plan must be applied from the date of commencement of and for the duration of the construction work and must be reviewed and updated by the principal constructor as work progresses; and*
- *create and keep an on-site H&S file which must include all documentation required in terms of the Act and these regulations. It must be made available on request to an inspector, the client, the client's agent, or a contractor.*

The rules and regulations of H&S are aimed at ensuring that workers remain safe. Nevertheless, poor enforcement and compliance by stakeholders plague South African construction as statistics of work-related accidents, injuries, disabilities, illnesses and fatalities continues to soar. Typical reasons responsible for poor compliance are the view by stakeholders that these regulations are unattainable, expensive and a burden that has an unnecessary cost implication (Windapo, 2013). It is safe to assume that this picture is prevalent in not only South Africa but other countries in SSA as well, as there is evidence of reports surrounding poor enforcement and compliance to H&S regulations from the region as well as globally (Fugas et al., 2012).

In Nigeria, there is poor H&S knowledge and low levels of management commitment to H&S which trickle down to contractors and their subcontractors, thereby making accident reporting very low (Umeokafor & Isaac, 2016). Again, factors such as the perceived financial cost of H&S implementation, poor H&S knowledge and a nonchalant attitude towards how organisations' non-compliance with H&S impacts their image rank high among a plethora of identified barriers to H&S self-regulation in Nigeria (Umeokafor, 2017).

1.3 Safety Climate in Construction Operations

Owing to the complexity and continuously evolving nature of the construction site, the need for a well-developed management system that emphasises H&S-related behaviour and attitude cannot be overemphasised (Whiteoak & Mohamed, 2016). The study of safety climate and culture is not unique to construction; it pertains to other high-risk work environments as well. This explains the constantly evolving literature and concepts aimed at the development of models or frameworks for the elimination or reduction to the barest minimum of workplace-related accidents. The definitions of both H&S climate and culture have continued to evolve over time; however, the major factor in safety climate is the employees' perceptions relative to safety-related policies, procedures and practices in the workplace (Alruqi et al., 2018). Safety culture, on the other hand, has equally enjoyed debates by scholars, especially what constitutes an acceptable definition when viewed from the standpoint of culture as a concept. Strauch (2015) acknowledged some of these definitions and concluded that safety culture may be viewed as employees' visible commitment and behaviour to the valuing of safe operations within the organisation as they go about their roles. According to Zohar (1980), management is solely responsible for the establishment and control of a safety climate. Management must build a safety climate that gives the workers the perception that H&S goals are equally as important as meeting deadlines, productivity targets and profit maximisation, among others (Dov, 2008).

In a literature network analysis based on 38 years of H&S research (1980–2019), Bamel et al. (2020) revealed that studies in safety climate have multiplied by 11 per cent annually, and newer methods such as longitudinal analysis, mediation and moderation analysis, as well as multilevel analysis are being employed by researchers in the examination of safety climate. Since safety climate reflects H&S perspectives shared by individuals in the same work environment, it implies that these perceptions may vary from person to person based on their understanding of what H&S means. This again may be determined by whose perceptions are being assessed. For example, a construction worker who occupies a management position may view H&S differently from a manual construction worker. Literature again emphasises that management's safety behaviour may predict that of workers. Management must demonstrate acceptable safety-related behaviour, given that power dynamics in their relationship with workers puts them in a position to act as mentors to workers.

As mentioned earlier, the concept of safety climate in literature is often discussed in tandem with safety culture; the likelihood of misunderstanding both concepts will naturally increase owing to limited insight. Reason (2003) broadly captured culture as a concept that surpasses an individual's mindset and can establish reminders and ways of working which will create a sustainable and intelligent reminder of how to remain within H&S confines in the workplace. Yet He et al. (2012) cited the Assessment Standards of Enterprise Safety Culture Developing's definition in their literature as 'the sharing of safety-related values, attitudes, ethics and code of conduct in an organisation which eventually provides a working direction for an effective safety culture construction'. People who grew up in a prevailing culture will naturally behave similarly. As every geographical location may have a set of accepted norms enshrined in their culture and one may differ from another, so this also applies to work organisations (Misnan & Mohammed, 2007).

Edwards et al. (2013) recognised over 100 definitions of safety culture found in literature and all agree on a commonality in safety-related attitudes and behaviour, especially in the workplace. Culture in the broad sense considers behaviour among a group of persons, as informed by similar values and beliefs often passed from a previous generation to the present and to future generations to ensure the preservation of such norms (Strauch, 2015). This implies that one organisation may have an H&S culture different from another. Cole et al. (2013) described an organisation's culture as a set of workable behaviours that are developed and applied to solving problems in the workplace. These methods, if considered valid by the organisation, are maintained and taught to new members as appropriate techniques to be applied to problem-solving (Schein, 2015; Health and Safety Executive, 2005). This implies that an organisation's culture is a representation of the values and beliefs it represents and practises.

Organisational climate and culture, as conceptualised by Guldenmund (2010), include climate as a generic term used in reference to broad and diverse dimensions in the organisation which eventually makes it useless and the concept of culture as a blank space that is hard to define, analyse, measure and manage. While this conceptualisation may appear vague, it implies that the concept of organisational climate is broader than culture and if culture is a blank space, it may again imply that a safety climate may define or influence the safety culture in the workplace. In general, safety climate is the collective perception of workers in an organisation which defines the workers' behaviour towards H&S. It is evident that management plays a leading role in determining the safety climate in the workplace and invariably sets in place a safety culture. Organisational safety may then refer to the organisation's acceptable norms, values, beliefs and approach to problem-solving with respect to H&S.

The earliest measure of an organisational climate was done by Zohar in 1980, whose factor analysis identified principal components in a safety climate to include perceived management attitudes on safety, the effect of safe work practice on promotion, the social status of the safety committee, the importance and effectiveness of safety training, risk in the workplace and enforcement versus guidance. The outcome revealed that an organisation's safety climate is linked to its safety record; an analysis of these records shows the organisational areas which need improvement (Kukoyi & Adebowale, 2021). Over the years, the components of the organisational climate identified by Zohar and the outcome of the study have received industry-wide criticism and review; for example, the measurement of a safety climate should include the employees' perceptions of safety-related conditions, which influence their motivation and the relational aspects of occupational H&S (Kines et al., 2011). Choudhry (2014) opined that when a group of people begin to share common attitudes and habits behaviours concerning H&S, a culture is formed.

Organisational climate is more concrete and easier to measure than culture is (Dov, 2008). Nevertheless, the values, and priority accorded to the organisational climate, such as bordering on H&S, stem from the core beliefs and values held by managers who can relate to a prevailing culture. This implies that the values held by management make up their culture and determine their own behaviour, which then influences workers' perception of safety climate. Safety climate and culture are intertwined and when subjected to time reveal their influence on workers' behaviour (Seo et al., 2015).

1.3.1 Construction Safety Performance

Safety performance is an organisational metric both for safety outcomes and for workers' compliance behaviour (Christian et al., 2009). Choudhry (2014) defined

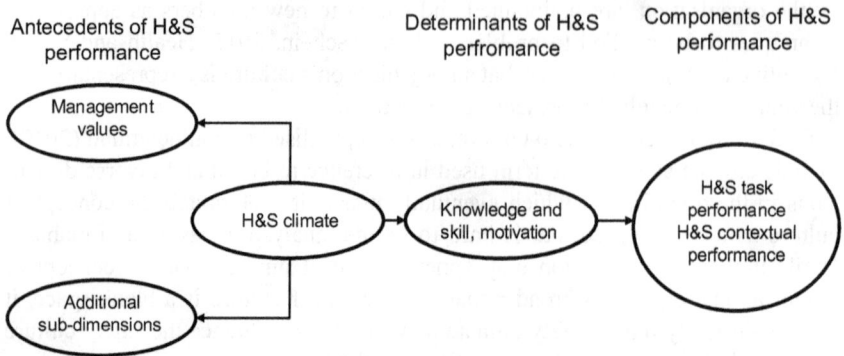

Figure 1.1 The relationship among antecedents, determinants and components of safety performance.

Source: Modified from Griffin and Neal (2000).

safety performance as the outcome of identifying and removing all or any factor that increases the likelihood of accident occurrence. Examples of these factors include counterproductive work behaviour (CWB) such as refusing to wear personal protective equipment (PPE) or an inability of management to enforce H&S compliance. Safety performance is thus the outcome of safety-related goals as set by management. This suggests the organisation's position with regard to safety compliance when assessed.

Despite the importance of safety performance to safety research, there seems to be an absence of a clear and consistent description or definition of the concept as it is often viewed as a factor of either compliance or participation (Christian et al., 2009). As rightly stated by Griffin and Neal (2000), safety performance may be influenced by three individual factors, namely knowledge, skill and motivation. They insisted that these factors determine safety performance since workers can only perform within the scope of their knowledge and skill and according to how motivated they are to engage in safety behaviour.

According to the Health and Safety Executive (2001), safety performance measurement is an important audit step which aids the management's safety-related decision-making in the workplace. The report further lists questions any H&S performance measurement must seek to answer:

- What is the organisation's current position in terms of its overall H&S objectives?
- Where is the organisation in terms of controlling hazards and risk?
- How does the organisation compare with others?
- Is the organisation's H&S performance improving or worsening over time?
- Is the management of H&S effective, reliable, efficient and proportionate to its hazards and risk?

- Is there an effective H&S management system across all parts of the organisation?
- Does the organisational culture support H&S especially in the face of other equally important and competing demands?

Answering these questions provides a holistic picture of the organisation's H&S performance and should be answered at all management levels.

1.3.2 Management Safety Values

Values are broadly referred to as 'operating philosophies or principles that guide the internal conduct of a group as well as their relationship with other groups. Values may inform a person's perception of what is good, desirable, or unacceptable, it may influence behaviour and serve as a broad guideline for situations' (Zwetsloot et al., 2013). Ideally, every organisation has a value statement that guides the management and business conduct. It is safe to assume that construction firms will consider compliance to H&S regulations in the workplace a core value. According to Halbesleben et al. (2013), management's behavioural integrity has a direct influence on employee compliance behaviour. For example, if management claims that H&S is at the core of its values, workers are very likely to comply on the grounds that they see behaviour which supports such a claim from management; it is a case of walking the talk. This further implies that providing the appropriate knowledge, training and equipment may communicate to the worker that management values H&S. Conversely, rewarding only productivity performance may signal to workers that H&S compliance behaviour is not considered a core value. Maierhofer et al. (2000) maintained that personal values may influence individual behaviour and may change from an organisational perspective. Organisations are conceptualised as strong institutions and workers may adopt or restructure their personal values to match organisational values in the workplace. This again implies that organisational values which have H&S compliance at their core may dictate or guide the values of the individuals who occupy the position of management towards an improved safety performance.

1.3.3 Safety Performance Indicators

Safety performance is often measured using metrics such as the recordable injury rate, days away from work, rate of injury and restricted work. Hinze et al. (2013), however, argue that these only provide historical information since they are post-incident and as such should be lagging indicators. H&S leading indicators are measures that can be used futuristically as predictors of H&S performance. Beus et al. (2016) added that H&S compliance behaviour in the workplace is considered an H&S performance indicator because it clearly shows patterns that can be curbed to prevent the occurrence of an accident while the accident is a lagging indicator. This implies that indices such as low accident occurrence in the workplace may not necessarily translate into an outstanding H&S performance rating as workers may be engaging in unsafe behaviour that has not yet resulted in an accident.

The effective monitoring of safety performance indicators will provide a data trail which may be employed by management as an instrument for setting developmental goals and assessing the efficacy of previous improvements in the workplace (Reiman & Pietikäinen, 2010). According to Alruqi and Hallowell (2019), leading indicators that have shown a strong correlation with worksite injuries include the following:

- Safety records
- Safety resources
- Staffing for safety
- Owner involvement
- Safety training or orientation
- PPE
- Safety incentives programme
- Safety inspection and observation
- Pre-task safety meetings

Sadly, literature on construction safety performance indicators still requires improvement; stakeholders such as researchers in the field of safety climate must actively contribute to the debate. It is important that safety performance indicators are measurable, targeted and can be used as a database to determine future indicators. Safety performance may be measured on a project-to-project basis. As a precursor, management may set safety goals which are standardised, measurable and attainable at the commencement of each project. These goals are then assessed at the end of the project to determine the organisation's safety performance on the project.

1.3.4 Barriers to Safety Performance

Despite the overwhelming evidence of global interest from researchers and professionals regarding issues which border on construction safety climate, debates concerning barriers to safety performance have improved in some climes. In regions such as SSA, there is still much that needs to be done in order to improve existing safety performance levels in the construction industry. The gap in safety performance observed between developed and developing countries includes inadequate management commitment, poor supervision and monitoring, visible difference in safety training and competencies, non-existent regulatory system and excessive subcontracting (Kukoyi & Adebowale, 2021). Literature from Sunindijo (2015) and Da Silva and Amaral (2019) suggests that barriers to the safety performance may include the following:

- **Organisational size**. Small contractors often offer lower cost of delivery. This may be related to the fact that they do not consider H&S compliance a priority and will aim at maximising their profit.

- **Multiple subcontractors on a project**. It is difficult to monitor H&S compliance on a project which has multiple subcontractors. Often, these subcontractors offer lower rates of delivery and though the main contractors may be compliant, they may also want to maximise profit by employing the services of a subcontractor who may not necessarily consider safety a priority or who has offered a lower bid because H&S was eliminated from their bid.
- **Client's priority**. Clients who do not understand the value of H&S or do not make its compliance a priority may not understand the cost of its implementation. This may inform their decision to opt for a lower bid from a contractor who did not consider the same at the procurement phase of the project. To the client, a contractor who offers them traditional deliverables such as time, cost and quality invariably satisfies their objectives.

The goal of the construction industry globally is to attain a zero-accident workplace. This feat, though not impossible, is a function of an improved safety leadership and management. Smallwood and Emuze (2016) confirmed that a zero fatality or a zero-accident workplace is highly attainable if management will value the people in construction by making H&S a project value as opposed to it being a contractor's responsibility. This can be achieved only if there is a partnership and synergy involving all project stakeholders. Improved H&S performance in the construction industry revolves around the various steps and modalities which must be put in place to achieve a workplace with zero accidents (Sanni-Anibire et al., 2020). This information can only be ascertained by an assessment of data retrieved from accident records and their various causes in an H&S performance evaluation, which provides information absent in incident-based measurements and tells the overall performance of the organisation's H&S management system in safe operation (Wu et al., 2015).

1.4 Management Safety Practice

The responsibility of ensuring that H&S regulations are adhered to lies with the management, given that it is their sole responsibility to set up policies which ensure that compliance is attained on the worksite. Mohammadi et al. (2018) stated that poor support from management contributes substantially to construction workers' inability to achieve set safety-related goals. Management's poor knowledge of H&S issues and an unflinching interest in driving workers towards achieving productivity-related goals for the maximisation of profit over their H&S have also been found to contribute to workers' non-compliant behaviour (Kapp, 2012). According to Emuze et al. (2016), the safety climate in an organisation is greatly influenced by management's vision, goal and beliefs, implying that if the management in an organisation fails to make workers' H&S a priority, workers may also not consider it as such. Though the responsibility for the effective implementation of the construction H&S programme is ideally shared between management and workers, the main responsibility is dependent on management's actions. Successful safety management requires the support of an organisation that displays its attitudes and behaviours towards the attainment of a strong H&S system (Griffith & Howarth, 2014).

The International Social Security Association's protocol for the integration of H&S competencies into vocational and technical education directly highlights the role of management in safety practice. Among other requirements, the protocol sets out four principles that actors involved in H&S training, which informs practice, should observe. These include the integration of H&S competencies into the education process required to perform a task, mastery of knowledge and practices related to H&S to get tasks completed, promotion of exemplary H&S practices through policies and rules and the use of equipment and facilities that meets required H&S standards (International Social Security Association, 2003). This is why clients, designers and contractors in construction cannot afford not to have the right policies in place.

In addition, although contractors may have a well thought out H&S policy in writing, implementation levels of these policies is often lacking. Furthermore, an effective H&S climate is only possible if such a written policy is implemented or enforced in favour of an improved H&S performance (Okorie, 2014). Forcing workers to memorise the construction H&S policies will only get them to accomplish their jobs and may not necessarily impact on their H&S compliance behaviour. Management must not only invest in construction H&S training but also invest in continually creating more awareness among their workforce, seeing that different job roles may require different H&S training and levels of awareness (Mohammadi et al., 2018).

Organisational behaviour, which is the study of an employee's behaviour outside the workplace, can be explored or harnessed for the benefit of meeting workplace-related objectives (Ivanko, 2013). On the other hand, organisational culture can be said to represent the value system and acceptable methods of problem-solving within the workplace which are taught to newcomers to ensure sustainability among both employers and employees (Schein, 2015; Naha et al., 2011). This implies that the employers may or may not have a separate set of behaviour from that of the employees. In a power relationship between management and employee, it can be safe to assume that the employer sets the example for what organisational values are acceptable and practised and these are retained by the employees. Organisational culture often predates the employee and is passed on from one to the other as accepted modus operandi (Ali et al., 2015; Strauch, 2015)

1.5 Construction Workers' Safety Behaviour

A strong or weak organisational climate does not necessarily determine the success or failure of the organisation (Ghinea & Brătianu, 2012). This means that workers may or may not be susceptible to change, irrespective of the organisational culture in place. Managers are saddled with the responsibility of enforcing organisational values that support their workforce in the attainment of the organisation's goals and objectives. Organisational climate in tandem with the internal and external environments naturally wins the employees to its side. The workers may behave in a manner which suggests that they are fully committed to the attainment of the organisations goals and objectives or behave in a manner that counters the attainment of the same in organisations. These behaviours are grouped as organisational citizenship behaviour (OCB) or CWB.

Accidents in construction are often related to unsafe acts. In a briefing by Manu et al. (2017), it was stated that latent failures must be mitigated to the barest minimum or preferably removed from construction sites in addressing unsafe behaviour among construction workers. A meta-analysis conducted by Khosravi et al. (2014) revealed the following factors as key to influencing construction workers' unsafe behaviour:

- **Individual characteristics**. One of the most important factors influencing unsafe behaviour and accidents is found to be individual characteristics. These are further categorised in seven themes: attitude and motivation; age and experience; drug abuse; unintended acts; intended acts; competency and ability and psychological distress.
- **Site conditions**. The construction activity often occurs in an evolving work environment and condition. This category covers an extended range of themes including hazardous operations, unsafe conditions, equipment, weather, welfare services and the stage of construction. Construction-related activities may be tagged as risky considering the factors which come into play, such as the complexities of site plant and machinery, workers' H&S behaviour and working at height.
- **Workgroup**. This describes a collection of workers who are most likely engaged in a similar work activity. The workgroup is characterised by group norms and attitudes, group interaction and teamwork. Group norms are said to be adopted attitudes and behaviour permitted among a group of people. When positive H&S behaviour and attitudes towards H&S are established within a workgroup, the management of H&S becomes less complex.
- **Supervision**. This factor is also further broken into six themes, namely effective enforcement, supervision style, H&S engagement, communication, competency and performance pressure.
- **Project management**. Six themes emerge, namely H&S leadership, management commitment and support, management style, H&S communication, competency and review and feedback. Management's commitment to H&S often goes a long way in terms of reducing accidents and boosting H&S performance.
- **Organisation**. This is an important factor which influences unsafe behaviour among workers, resulting in accidents and near misses. This may further be divided into seven dominant themes, namely policy and plan, H&S climate and culture, structure and responsibility, information management, project and job design, contract and resource management.
- **Social factors**. The effect of social factors on unsafe behaviour and accidents might seem negligent from a distance but cannot be overemphasised. This factor is also divided into a further seven themes: societal culture, race or

ethnicity, education and knowledge, economy, social support, social challenges and business climate. Typical examples may include national culture, local worker and cultural and language problems, social support, race and ethnicity. In a market-driven society, it is common for construction stakeholders at the lower end of the supply chain to focus on the completion of projects at a specified cost, time and quality, thereby making issues surrounding construction H&S less of a priority, which often results in grave consequences.

1.6 H&S Compliance Issues in Developing Countries

In many countries in SSA, issues which border around compliance to H&S regulations in high-risk workplaces, such as construction, remain neglected. Though levels of neglect may vary among countries, it is established that globally, construction workers in developing countries have an 80 per cent likelihood of exposure to work conditions which increase the likelihood of accident occurring compared to workers in developed countries (Khan, 2013). This implies that a construction worker in Ghana or Kenya is more likely to be involved in a construction accident than one in Denmark or France. Safety compliance is the act of adhering to statutory legislations and regulations which have been set and passed to ensure that workers' H&S in the workplace is protected and preserved (Zin & Ismail, 2012). Though the list of challenges which inhibit compliance to H&S regulations in SSA countries can be inexhaustive, certain factors have consistently appeared in literature and outcomes from empirical studies. These include the following:

- **Non-existent OHS regulations**. There is mostly an absence of any regulation which addresses OHS and workers' rights. In instances where this is found, it is extracted from a developed country and often obsolete. Very few countries show evidence of having an OHS regulation which addresses their country's unique peculiarities and people.
- **Low levels of enforcement of set regulations**. Countries which have legislation passed in favour of OHS compliance have bemoaned the poor enforcement of these regulations in practice. In some countries, workers are trained to memorise some regulations; however, merely memorising these is hardly reflected in workers' safety-related behaviour.
- **Self-regulating by contractors**. Owing to the absence of an OHS regulation or poor levels of enforcement in some counties, contractors often self-regulate. This implies that the implementation of H&S regulation may be limited by the contractors' knowledge of the same and carried out according to their dictate. In such instances, the outcomes may be grossly undesirable.
- **Low levels of construction H&S knowledge**. This factor is prevalent in almost every category of construction worker. Low levels of H&S knowledge imply that workers and management are ignorant of the implication of non-compliance. This scenario may be disastrous for all parties involved because they will go about their duties in blissful ignorance and then blame avoidable accidents on natural causes.

- **Low levels of construction H&S training and retraining of workers**. Knowledge is not static and demands that management and workers update their H&S knowledge continuously. Contractors often avoid training and retraining of workers, owing to the cost implication involved. However, the cost of this process must not detract from its importance.
- **Non-consideration of H&S compliance as an equally important priority**. Both clients and contractors consider project goals such as timely completion, quality of output and cost-effectiveness to be a constant priority. Researchers have unintentionally skirted over these three objectives when discussing the importance of H&S-related outcomes. It is important to note that all objectives are paramount to the attainment of an improved safety climate and culture. An acceptable set of objectives for both client and contractors must be timely completion, quality of output, cost-effectiveness and safety performance.
- **Low levels of safety leadership and communication among construction workers**. Management must understand that their position on H&S compliance describes the type of leadership they represent. An effective leadership style is sustained by an equally effective communication system established in the workplace. Poor communication of H&S-related requirements may inhibit its compliance irrespective of how well-crafted the leadership approaches H&S.
- **Unsafe working environment and conditions**. Workers are often blamed for behaviour which does not support H&S compliance in the workplace. It is important that the conditions which allow non-compliance to H&S regulations to remain unchecked are equally addressed.

1.7 Conclusion

This chapter presents an overview of the safety climate and some challenges it faces. Safety climate thrives when both compliance-based safety and behaviour-based safety are visible in the workplace. Removing barriers that disempower operatives enhances safety climate. Such worksites uphold better safety performance by tackling the factors that discourage trust, for example. Generally, front-line construction workers may have low levels of formal education which may further influence their H&S knowledge and safety compliance. Since safety culture may also determine management safety-related behaviour, a careful understanding of the unique cultures must be studied in the development of a safety compliance approach to improve safety performance outcomes.

References

Ali NM, Jangga R, Ismai M, et al. (2015) Influence of leadership styles in creating quality work culture. *Procedia Economics and Finance* 31(15). Elsevier BV: 161–169. http://doi.org/10.1016/S2212-5671(15)01143-0

Alkanam N and Afatsawu PK (2022) A study of challenges faced by regulatory authorities for implementing health and safety compliance in the Ghana construction industry context. *International Journal of Management & Entrepreneurship Research* 4(7). Fair East Publishers: 315–333. http://doi.org/10.51594/ijmer.v4i7.353

Alruqi WM and Hallowell MR (2019) Critical success factors for construction safety: Review and meta-analysis of safety leading indicators. *Journal of Construction Engineering and Management* 145(3): 1–10. http://doi.org/10.1061/(ASCE)CO.1943-7862.0001626

Alruqi WM, Hallowell MR and Techera U (2018) Safety climate dimensions and their relationship to construction safety performance: A meta-analytic review. *Safety Science* 109(June). Elsevier: 165–173. http://doi.org/10.1016/j.ssci.2018.05.019

Amponsah-Tawiah K and Mensah J (2016) Occupational health and safety and organizational commitment: Evidence from the Ghanaian mining industry. *Safety and Health at Work* 7(3). Elsevier: 225–230. http://doi.org/10.1016/j.shaw.2016.01.002

Asanka WA and Ranasinghe M (2015) Study on the impact of accident on construction project. In: *6th international conference on structural engineering and construction management 2015*, pp. 58–67. Available from: www.mom.gov.sg/

Awwad R, El Souki O and Jabbour M (2016) Construction safety practices and challenges in a Middle Eastern developing country. *Safety Science* 83. Elsevier: 1–11. http://doi.org/10.1016/j.ssci.2015.10.016

Bamel UK, Pandey R and Gupta A (2020) Safety climate: Systematic literature network analysis of 38 years (1980–2018) of research. *Accident Analysis and Prevention* 135(November 2019). Elsevier: 105387. http://doi.org/10.1016/j.aap.2019.105387

Beus JM, Mccord MA and Zohar D (2016) Workplace safety: A review and research synthesis. *Organisational Psychology Review* 6(4). SAGE Publishing: 352–381. http://doi.org/10.1177/2041386615626243

Boadu EF, Wang CC and Sunindijo RY (2021) Challenges for occupational health and safety enforcement in the construction industry in Ghana. *Construction Economics and Building* 21(1). UTS ePress: 1–21. http://doi.org/10.5130/AJCEB.v21i1.7482

Cheng CW, Leu S-S, Lin CC, et al. (2010) Characteristic analysis of occupational accidents at small construction enterprises. *Safety Science* 48. Elsevier: 698–707. http://doi.org/10.1016/j.ssci.2010.02.001

Chim BLH, Chun CJ and Wah FK (2018) Accidents in construction sites: A study on the causes and preventive approaches to mitigate accident rate. *INTI Journal* 1(3): 1–12. Available from: http://eprints.intimal.edu.my/1136/

Choudhry RM (2014) Behavior-based safety on construction sites: A case study. *Accident Analysis and Prevention* 70. Elsevier: 14–23. http://doi.org/10.1016/j.aap.2014.03.007

Christian MS, Bradley JC, Wallace JC, et al. (2009) Workplace safety: A meta-analysis of the roles of person and situation factors. *Journal of Applied Psychology* 94(5). American Psychological Association: 1103–1127. http://doi.org/10.1037/a0016172

Cole KS, Stevens-Adams SM and Wenner CA (2013) *A literature review of safety culture*. Albuquerque, New Mexico: Sandia National Laboratories. Available from: https://www.osti.gov/servlets/purl/1095959

Dane E and Brummel BJ (2014) Examining workplace mindfulness and its relations to job performance and turnover intention. *Human Relations* 67(1). Sage Publications: 105–128. http://doi.org/10.1177/0018726713487753

Da Silva SLC and Amaral FG (2019) Critical factors of success and barriers to the implementation of occupational health and safety management systems: A systematic review of literature. *Safety Science* 117(February). Elsevier: 123–132. http://doi.org/10.1016/j.ssci.2019.03.026

Dov Z (2008) Safety climate and beyond: A multi-level multi-climate framework. *Safety Science* 46(3). Elsevier: 376–387. http://doi.org/10.1016/j.ssci.2007.03.006

Edwards JRD, Davey J and Armstrong K (2013) Returning to the roots of culture: A review and re-conceptualisation of safety culture. *Safety Science* 55. Elsevier: 70–80. http://doi.org/10.1016/j.ssci.2013.01.004

Emuze F (2018) A descriptive study of human errors producing unsafe acts in construction. In: *Proceedings of the joint CIB W099 and TG59 conference coping with the complexity of safety, health, and wellbeing in construction*, Salvador, Brazil, pp. 310–317.

Emuze F, Linake M and Seboka L (2016) Construction work and the housekeeping. In: *Proceedings of the 32nd annual ARCOM conference* 1(September), Association of Researchers in Construction Management, pp. 537–546.

Fang W, Love PED, Luo H, et al. (2020) Computer vision for behaviour-based safety in construction: A review and future directions. *Advanced Engineering Informatics* 43(August 2019). Elsevier: 100980. http://doi.org/10.1016/j.aei.2019.100980

Fu G, Xie X, Jia Q, et al. (2020) The development history of accident causation models in the past 100 years: 24Model, a more modern accident causation model. *Process Safety and Environmental Protection* 134(February 2020). Elsevier: 47–82. http://doi.org/10.1016/j.psep.2019.11.027

Fugas CS, Silva SA and Meliá JL (2012) Another look at safety climate and safety behavior: Deepening the cognitive and social mediator mechanisms. *Accident Analysis and Prevention* 45. Elsevier: 468–477. http://doi.org/10.1016/j.aap.2011.08.013

Ghinea VM and Brătianu C (2012) Organizational culture modeling. *Management and Marketing Challenges for the Knowledge Society* 7(2). Society for Business Excellence: 257–276.

Gibb A, Lingard H, Behm M, et al. (2014) Construction accident causality: Learning from different countries and differing consequences. *Construction Management and Economics* 32(5). Routledge: 446–459. http://doi.org/10.1080/01446193.2014.907498

Griffin MA and Neal A (2000) Perceptions of safety at work: A framework for linking safety climate to safety performance, knowledge, and motivation. *Journal of Occupational Health Psychology* 5(3). American Psychological Association: 347–358. http://doi.org/10.1037/1076-8998.5.3.347

Griffith A and Howarth T (2014) The nature of construction health and safety. In: *Construction health and safety management*. New York: Routledge, pp. 1–10.

Guldenmund FW (2010) *Understanding and exploring safety culture*. Oisterwijk: Uitgeverij BOXPress Postbus. http://doi.org/10.1111/j.1539-6924.2010.01452.x

Guo BHW, Yiu TW and González VA (2016) Predicting safety behavior in the construction industry: Development and test of an integrative model. *Safety Science* 84. Elsevier: 1–11. http://doi.org/10.1016/j.ssci.2015.11.020

Halbesleben JRB, Leroy H, Dierynck B, et al. (2013) Living up to safety values in health care: The effect of leader behavioral integrity on occupational safety. *Journal of Occupational Health Psychology* 18(4). American Psychological Association: 395–405. http://doi.org/10.1037/a0034086

He A, Xu S and Fu G (2012) Study on the basic problems of safety culture. *Procedia Engineering* 43. Elsevier: 245–249. http://doi.org/10.1016/j.proeng.2012.08.042

Health and Safety Executive (2001) *A guide to measuring health & safety performance*. Suffolk: HSE Books. Available from: https://safetyresourcesblog.files.wordpress.com/2014/09/a-guide-to-measuraing-health-and-safety.pdf

Health and Safety Executive (2005) *A review of safety culture and safety climate literature for the development of the safety culture inspection toolkit*. Sulfolk: HSE Books. Available from: www.hse.gov.uk/research/rrpdf/rr367.pdf

Hinze J, Hallowell M and Baud K (2013) Construction – safety best practices and relationships to safety performance. *Journal of Construction Engineering and Management* 139(10): 04013006. http://doi.org/10.1061/(ASCE)CO.1943-7862.0000751

Hosseinian SS and Torghabeh ZJ (2012) Major theories of construction accident causation models: A literature review. *International Journal of Advances in Engineering and Technology* 4(2). IJAET: 53–66.

International Social Security Association (2003) *Quebec city protocol for the integration of occupational health and safety (OHS) competencies into vocational and technical education.* Available from: ww1.issa.int/sites/default/files/documents/prevention/2_-_Quebec_City_Protocol_en-36594.pdf (accessed 17 October 2022).

Ivanko Š (2013) *Organizational behaviour.* Ljubljana: University of Ljubljana Faculty of Public Administration.

Kapp EA (2012) The influence of supervisor leadership practices and perceived group safety climate on employee safety performance. *Safety Science* 50(4). Elsevier: 1119–1124. http://doi.org/10.1016/j.ssci.2011.11.011

Khan MI (2013) Developing a safety culture in the healthcare workplace. In: *International conference on safety, construction engineering and project management*, Islamabad, Pakistan, pp. 26–31.

Khosravi Y, Asilian-Mahabadi H, Hajizadeh E, et al. (2014) Factors influencing unsafe behaviors and accidents on construction sites: A review. *International Journal of Occupational Safety and Ergonomics* 20(1). Central Institute for Labour Protection: 111–125. http://doi.org/10.1080/10803548.2014.11077023

Kines P, Lappalainen J, Mikkelsen KL, et al. (2011) Nordic Safety Climate Questionnaire (NOSACQ-50): A new tool for diagnosing occupational safety climate. *International Journal of Industrial Ergonomics* 41(6). Elsevier: 634–646. http://doi.org/10.1016/j.ergon.2011.08.004

Kukoyi PO and Adebowale OJ (2021) Impediments to construction safety improvement. *Journal of Engineering, Project, and Production Management* 11(3). EPPM Association: 207–214. http://doi.org/10.2478/jeppm-2021-0020

Lestari F, Sunindijo RY, Loosemore M, et al. (2020) A safety climate framework for improving health and safety in the Indonesian construction industry. *International Journal of Environmental Research and Public Health* 17(20). MDPI: 1–20. http://doi.org/10.3390/ijerph17207462

Liang H and Zhang S (2019) Impact of supervisors' safety violations on an individual worker within a construction crew. *Safety Science* 120. Elsevier: 679–697. http://doi.org/10.1016/j.ssci.2019.08.014

Maierhofer NI, Griffin MA and Sheehan M (2000) Linking manager values and behavior with employee values and behavior: A study of values and safety in the hairdressing industry. *Journal of Occupational Health Psychology* 5(4). American Psychological Association: 417–427. http://doi.org/10.1037//1076-8998.5.4.417

Manu P, Gibb A, Manu E, et al. (2017) Briefing: The role of human values in behavioural safety. *Proceedings of Institution of Civil Engineers: Management, Procurement and Law* 170(2). ICE Publishing Ltd.: 49–51. http://doi.org/10.1680/jmapl.16.00047

Misnan MS and Mohammed AH (2007) Development of safety culture in the construction industry: A conceptual framework. In: Boyd D (Ed.), *Procs 23rd annual ARCOM conference*, Belfast, UK, Association of Researchers in Construction Management, 3–5 September 2007, pp. 13–22. Available from: https://www.arcom.ac.uk/-docs/proceedings/ar2007-0013-0022_Misnan_and_Mohammed.pdf

Mohammadi A, Tavakolan M and Khosravi Y (2018) Factors influencing safety performance on construction projects: A review. *Safety Science* 109(November 2018). Elsevier BV: 382–397. http://doi.org/10.1016/j.ssci.2018.06.017

Naha N, Mansur A, Ahmed MA, et al. (2011) Personality and organizational outcomes (organizational culture as a moderator). *International Journal of Academic Research* 3: 54–59. Available from: www.ijar.lit.az

Okonkwo PN (2019) *Health and safety management and performance among construction contractors in South Africa.* PhD thesis, Stellenbosch University, Stellenbosch, South Africa.

Okorie V (2014) *Behaviour-based health and safety management in construction: A leadership-focused approach.* PhD thesis, Nelson Mandela Metropolitan University, Port Elizabeth, South Africa.

Olutende M, Wamukoya EK, Wanzala M, et al. (2021) Predictors of occupational health and safety management practices in the building construction industry, Kakamega Kenya. *Journal of Nursing and Health Services* 10(2). OJS/PKP: 43–57. http://doi.org/10.9790/1959-1002044357

Reason J (2003) Safety paradoxes and safety culture. *Injury Control and Safety Promotion* 7(1). Informa UK: 3–14. http://doi.org/10.1076/1566-0974(200003)7:1;1-v;ft003

Reiman T and Pietikäinen E (2010) *Indicators of safety culture selection and utilization of leading performance indicators (Report No. 2010:07).* Helsinki: VTT Technical Research Centre of Finland. Available from: https://publications.vtt.fi/julkaisut/muut/2010/SSM-Rapport-2010-07.pdf

Sanni-Anibire MO, Mahmoud AS, Hassanain MA, et al. (2020) A risk assessment approach for enhancing construction safety performance. *Safety Science* 121(September 2019). Elsevier: 15–29. http://doi.org/10.1016/j.ssci.2019.08.044

Schein EH (2015) *Organizational culture and leadership.* 3rd ed. San Francisco, CA: Jossey-Bass.

Seo HC, Lee YS, Kim JJ, et al. (2015) Analyzing safety behaviors of temporary construction workers using structural equation modeling. *Safety Science* 77. Elsevier: 160–168. http://doi.org/10.1016/j.ssci.2015.03.010

Smallwood J and Emuze F (2016) Towards zero fatalities, injuries, and disease in construction. *Procedia Engineering* 164. Elsevier BV: 453–460. http://doi.org/10.1016/j.proeng.2016.11.644

South African Government (2014) Construction Regulations. Government Gazette No 27305. Pretoria: Republic of South Africa.

Strauch B (2015) Can we examine safety culture in accident investigations, or should we? *Safety Science* 77. Elsevier: 102–111. http://doi.org/10.1016/j.ssci.2015.03.020

Sunindijo RY (2015) Improving safety among small organisations in the construction industry: Key barriers and improvement strategies. *Procedia Engineering* 125. Elsevier BV: 109–116. http://doi.org/10.1016/j.proeng.2015.11.017

Suraya I, Arzahan N, Ismail Z, et al. (2022) Safety culture, safety climate, and safety performance in healthcare facilities: A systematic review. *Safety Science* 147. Elsevier: 105624. http://doi.org/10.1016/j.ssci.2021.105624

Umeokafor NI (2017) Barriers to construction health and safety self-regulation: A scoping case of Nigeria. *Civil Engineering Dimension* 19(1). CED: 44–53. http://doi.org/10.9744/ced.19.1.44-53

Umeokafor NI (2018) Construction health and safety research in Nigeria: Towards a sustainable future. In: *Proceedings of the joint CIB W099 and TG59 conference coping with the complexity of safety, health, and wellbeing in construction*, Salvador, Brazil, 1–3 August 2018, pp. 213–221. Available from: https://www.irbnet.de/daten/iconda/CIB_DC31527.pdf

Umeokafor NI and Isaac D (2016) Construction health and safety self-regulation in developing countries: A Nigeria case study. *Journal for the Advancement of Performance Information and Value* 8(1). Kashiwagi Solution Model Inc.: 74–87. http://doi.org/10.37265/japiv.v8i1.45

Umeokafor NI, Windapo AO and Manu P (2018) Country context-based opportunities for improving health and safety. In: *Proceedings of the joint CIB W099 and TG59 conference coping with the complexity of safety, health, and wellbeing in construction*, Salvador, Brazil, 1–3 August 2018, pp. 177–186. Available from: https://www.irbnet.de/daten/iconda/CIB_DC31523.pdf

Whiteoak JW and Mohamed S (2016) Employee engagement, boredom and frontline construction workers feeling safe in their workplace. *Accident Analysis and Prevention* 93. Elsevier: 291–298. http://doi.org/10.1016/j.aap.2015.11.001

Windapo A (2013) Relationship between degree of risk, cost and level of compliance to occupational health and safety regulations in construction. *Australasian Journal of Construction Economics and Building* 13(2). UTS ePress: 67–82. http://doi.org/10.5130/ajceb.v13i2.3270

Windapo A and Oladapo A (2012) Determinants of construction firms' compliance with health and safety regulations in South Africa. In: *Proceedings 28th annual ARCOM conference*, Edinburg, UK, Association of Researchers in Construction Management, 3–5 September 2012, pp. 433–444. Available from: http://www.arcom.ac.uk/-docs/proceedings/ar2012-0433-0444_Windapo_Oladapo.pdf

Winge S, Albrechtsen E and Mostue BA (2019) Causal factors and connections in construction accidents. *Safety Science* 112(August 2018). Elsevier: 130–141. http://doi.org/10.1016/j.ssci.2018.10.015

Wu X, Liu Q, Zhang L, et al. (2015) Prospective safety performance evaluation on construction sites. *Accident Analysis and Prevention* 78. Elsevier: 58–72. http://doi.org/10.1016/j.aap.2015.02.003

Yang H, Chew DAS, Wu W, et al. (2012) Design and implementation of an identification system in construction site safety for proactive accident prevention. *Accident Analysis and Prevention* 48. Elsevier: 193–203. http://doi.org/10.1016/j.aap.2011.06.017

Yorio PL, Edwards J and Hoeneveld D (2019) Safety culture across cultures. *Safety Science* 120(July). Elsevier: 402–410. http://doi.org/10.1016/j.ssci.2019.07.021

Zin SM and Ismail F (2012) Employers' behavioural safety compliance factors toward occupational, safety and health improvement in the construction industry. *Procedia – Social and Behavioral Sciences* 36. Elsevier BV: 742–751. http://doi.org/10.1016/j.sbspro.2012.03.081

Zohar D (1980) Safety climate in industrial organizations: Theoretical and applied implications. *Journal of Applied Psychology* 65(1). American Psychological Association: 96–102. http://doi.org/10.1037/0021-9010.65.1.96

Zwetsloot GIJM, Kines P, Ruotsala R, et al. (2017) The importance of commitment, communication, culture and learning for the implementation of the Zero Accident Vision in 27 companies in Europe. *Safety Science* 96. Elsevier: 22–32. http://doi.org/10.1016/j.ssci.2017.03.001

Zwetsloot GIJM, Scheppingen ARV, Bos EH, et al. (2013) The core values that support health, safety, and well-being at work. *Safety and Health at Work* 4(4). Elsevier: 187–196. http://doi.org/10.1016/j.shaw.2013.10.001

2 Safety Priority

2.1 Introduction

The influence of management's H&S behaviour on their employees will either promote or constrain the attainment of workplace safety-related goals. Safety priority in the workplace suggests that workers perceive that their H&S is regarded as a high priority regardless of other equally competing demands such as work productivity and speed of completion (Bosak et al., 2013). This is often displayed in the quality of safety-related support that frontline workers enjoy from management (Mohammadi et al., 2018). In any organisation, the management arm refers to the decision-making and implementing body, thus implying that they are not only saddled with the responsibility of setting goals for the organisation, but they are also setting up modalities which ensure that the set goals are achieved. The management arm represents leadership, which normally includes top-, mid- and low-level managers. At all levels, they play a significant role regarding employees H&S and may exhibit this by ensuring that a high level of safety performance is achieved in the workplace (Cheng et al., 2010).

In work environments such as construction, it is the responsibility of management to set up organisational policies that are in line with H&S regulations and ensure workers' compliance (Zohar & Luria, 2005). Okorie (2014) rightly observed that though contractors have a well thought out H&S policy in writing, the implementation of these policies is grossly lacking. The attainment of an effective safety climate is only possible if the written policy is implemented or enforced. Okorie added that the memorisation of construction H&S policies by workers will only get them to accomplish their jobs but does not ensure their compliance. In this chapter, safety priority is explored from the position of management and how their visible safety-related behaviour influences frontline workers' adoption of safety behaviour.

The chapter also presents empirical findings from data collected from construction professionals in two SSA countries, namely Nigeria and South Africa. The data were collected using a mixed-method approach and the instrument adopted for collecting quantitative data was the Management Safety Priority dimension on the Nordic Safety Climate Questionnaire (NOSACQ-50). Qualitative data were collected in another round of survey using open-ended questions where participants

DOI: 10.1201/9781003361640-2

enjoyed full anonymity and freely expressed themselves without having to be concerned that they were saying what they perceived the researcher needed them to say. Quantitative data were analysed using statistical tools and software such as SPSS while qualitative data were cleaned and assessed thematically.

2.2 Management Safety Commitment and Practice

Management may be aware of the concept of safety priority and may believe that they consider safety as high priority in the workplace; however, evidence reveals that this is not backed by a commensurate behaviour supporting the assumption (Liang & Zhang, 2019; Delegach et al., 2017). This may suggest that management's safety prioritisation, especially in SSA countries, needs to be improved. Safety priority ensures that management does not treat the H&S of frontline construction workers with an attitude which either subtly or overtly suggests that they do not matter, can be replaced easily or are dispensable. High poverty levels and acute job shortages in SSA may be the reason why many frontline workers in the region only find jobs in the construction industry, given that the manual activity requires a low level of formal education. Irrespective of the reason which has driven frontline workers into construction, management must be seen to devise appropriate methods that demonstrate and ensure an acceptable level of safety priority for the overall benefit of H&S compliance among workers. Windapo (2013) rightly observed that contractors (management) often consider issues relating to safety compliance as an unnecessary cost and they do everything possible to either avoid it or deliver below expectation if they perceive that they can get away with it. In SSA where issues which border on weak enforcement of safety regulations have been identified as a major factor inhibiting safety performance, contractors find it easy to manoeuvre their way around H&S compliance (Kheni & Braimah, 2014).

The responsibility for ensuring that H&S standards are adhered to lies with the management, considering that it is their duty to set up policies according to H&S regulations to ensure improved compliance levels to H&S regulations on the worksite. Mohammadi et al. (2018) observe that poor management support contributes substantially to construction workers' inability to achieve set H&S goals. Management's poor knowledge of H&S issues and a relentless pursuit towards achieving productivity-related goals for the maximisation of profit over their H&S also contribute to workers' H&S non-compliant behaviour (Kapp, 2012). The safety climate in an organisation is often significantly influenced by management's vision, goals and beliefs; if management in an organisation fails to make workers' H&S a priority, workers may not consider it as such either (Emuze et al., 2016). Though the responsibility for the effective implementation of the construction H&S programme is ideally shared among management and workers, the main responsibility lies with the management. Moreover, successful H&S management requires the support of an organisation that displays its attitudes and behaviours towards the attainment of a strong H&S system (Griffith & Howarth, 2014). Management must not only invest in construction H&S training but also invest in continually creating more awareness among their workforce and acknowledging that different job roles may require different H&S training and awareness (Mohammadi et al., 2018).

An organisational culture instilled by management represents the value system and acceptable methods of problem-solving in the workplace. This is taught to new employees to ensure sustainability across hierarchies (Schein, 2015; Mansur et al., 2011). Behaviour is defined by a prevailing culture, which Strauch (2015) described broadly as shared values, attitudes, beliefs and behaviour peculiar to a social group. These attributes are often passed down from generation to generation. Ali et al. (2015) corroborate this view, adding that the prevailing culture in an organisation influences the employees' views of where they work, their understanding of what they do and their perception of being part of the organisation. Therefore, the prevailing organisational culture in a workgroup often predates the workers, and the employer may determine what values are considered acceptable and are to be expressed and sustained in the organisation. The quality of an organisation's culture can be seen in how supportive it is of the development of its workers. For an organisation to be termed as efficient, it must have clear goals and objectives with a high standard of what they attribute to excellence, employee training and effective leadership (Ali et al., 2015). In an accident-prone work environment such as construction, it is imperative that management displays safety-related behaviour which suggests to workers that issues which relate to their H&S enjoy top priority and are not to be compromised. Behaviour is a strong tool which can be explored in workplaces such as construction to improve safety compliance irrespective of the workers' work level. Management must do everything within its power to ensure that H&S-related behaviour is enshrined in the organisation's culture.

2.2.1 Managements' Visible Safety Priority

Visible safety behaviour from management lies at the core of safety climate studies and a core predictor of work-related injuries (Alruqi et al., 2018). It is natural for employees (workers) to consider their employers (management) as mentors and will gradually imitate behaviour from senior colleagues. An organisation which prioritises safety eventually sets the example for the workers to follow in like manner. In regions such as SSA, a large proportion of workers in the construction industry have low levels of formal education; this single factor may impact on their self-esteem in work environments. Though this scenario may be deemed generally unfortunate, given that access to quality education is a fundamental human right, it is one which management may explore to influence workers by visibly showing that workers H&S is a priority not only in theory but also in practice. Low levels of formal education among frontline workers may imply that they may readily learn and imitate their leaders whom they regard as role models.

When management rewards or condemns certain behaviour exhibited by frontline workers, an immediate level of priority is attached to the reward or condemnation by the workers. Compliance to H&S regulations may be instinctive or extinctive and workers are likely to invest their energy in activities they perceive will be rewarded or recognised. According to Emuze et al. (2016), safety climate is largely influenced by the management's vision, goal and beliefs, implying that workers will prioritise their H&S based on what they glean from management. According to Choudhry and Fang (2008), management participation in H&S activities is the most effective factor for workplace H&S compliance. This is because workers believe that their H&S is the sole responsibility of management, thereby making predicting safety-related behaviour among workers a challenge for management and researchers. A positive safety priority being displayed from top management also trickles down to lower management officers such as supervisors. However, this category of management (supervisors) also has the capacity to thwart the efforts of management in the attainment of an improved safety priority, as they have been known to prioritise production pressures consistently over safety compliance (Newaz et al., 2019; Zohar & Luria, 2005).

2.3 Organisational Safety Policy

It is expected that the legislative arm of the government should pass laws which protect the OHS of all workers, especially in accident-prone workplaces, and also ensure that these laws are enforced and complied with. Locally at the organisational level, workplaces which consider safety a priority should have an H&S policy in place which is informed or driven by their values and which supports the organisation in attaining their safety-related goals and objectives. According to Lingard et al. (2012), large organisations need to pay heed to safety perceptions at the workgroup level as the organisation's size may sometimes make it difficult for top management to have contact with workers.

This implies that some workers may only know their supervisors and workgroup leaders as the representation of the organisation's safety priorities. This again emphasises the importance of safety supervisors and workgroup leaders in organisational safety priority representation. As earlier established, most frontline construction workers in SSA have low levels of formal education and are very likely to have joined the construction organisation with zero or minimal related work experience. This implies that they learn on the job through peer mentorship or by following instructions from supervisors or workgroup leaders. These two officers inevitably model organisational safety priorities to new workers, which may be used to sustain the organisational safety climate. Liang and Zhang (2019) observed that supervisors in the construction industry are often under pressure to meet productivity demands and, as a result, are the first to violate safety regulations by allowing workers to compromise. For example, insisting that workers wear the appropriate PPE may be overlooked by the supervisor if the workers claim that such PPE slows their ability to complete the task in a timely manner. An organisational safety policy may be tailored to address safety needs per project; in the event of an accident occurrence, the organisation must first identify its role in creating an environment which decreases the likelihood of incidents' and hazards' occurrence.

Organisational accidents, according to Strauch (2015), are those generally caused by initial conditions which have an impact on the immediate causes of the accident. Examples of these factors are recurrent factors which stem from a weak organisational culture and a complex or inappropriate organisation policy which does not have a feedback mechanism from operations and production pressures. This is another factor which may impact on the prevailing organisational reporting culture and workers' safety voice. Workers participate in an organisational safety policy when their views are sought in the process of drafting the organisation's safety policy (Wu et al., 2010). It is common expectation that the enforcement of OSH regulations and policies lies with a statutory authority. Lestari et al. (2020), however, noted that the drafting of such policies should be a collective responsibility, implying that all parties involved have a perfect understanding of what it entails and what the implication of non-compliance are. This may be challenging in workplaces with low levels of H&S knowledge; however, it is possible and worth attaining. Organisational safety policies must be clear and demonstrate that the H&S of all workers is considered an equal priority along with other competing objectives.

2.4 Workers' Perceived Safety Priority and Commitment

It is easy for workers to interpret an evident commitment to safety compliance from management as a safety priority. This perception increases their tendency to comply with H&S regulations as well, though with an expectation of being rewarded (Kines et al., 2011). Simply acknowledging workers' safety compliance may be all the reward they expect. Visible H&S compliance by top- and mid-level management, especially supervisors and workgroup leaders, has been established to be very effective in motivating workers to engage in the same. It is easy to get top- and mid-level management to engage in H&S compliance because they are likely to understand the implications of non-compliance and because they do not necessarily engage in manual construction work.

In some circumstances, management may be commended for upholding H&S regulations. It is important that they (management) understand such commendation is measured and assessed on the basis of the workers' compliance. The same factors which motivate workers towards safety compliance applies to frontline workers; for example, a supervisor who simply cheers workers who work safely will be seen as encouraging safety-related behaviour over one who scolds them for non-compliance. Positive affirmation from supervisors to workers may be easily interpreted as friendliness and mutual respect and makes it easier for supervisors to convince workers in favour of H&S compliance. Another approach management may adopt in motivating workers towards an improved H&S compliance when dealing with frontline construction workers in SSA is publicly acknowledging their compliance among their peers and colleagues; for example, management may organise an end-of-project meeting where workers give feedback on the just concluded project. Management may use this forum to acknowledge workers who displayed safety priority on the project. This step will encourage other workers to see that they receive similar awards on the next project and motivate those who were recognised to strive to maintain the same on another project.

Safety priority from workers is also influenced by co-workers (peers). This sometimes does not have anything to do with a colleague's rank or position in the organisation; it may simply be a case of one bad egg spoiling an entire crate, especially if such behaviour is condoned by management (Jiang et al., 2010). Conversely, workers with high safety compliance levels may also motivate their peers towards the same. An ideal workgroup is one which displays safety priority and where colleagues look out for one another on safety-related issues. Schwatka and Rosecrance (2016) found that management is able to influence workers towards H&S compliance when co-workers consider it a priority. This can be done through consistent safety training and toolbox talks. A healthy social exchange among workers rooted in strong safety communication for the benefit of safety compliance must be encouraged in the workplace.

2.4.1 Compliance and Commitment-Based Safety Compliance Among Workers

Construction workgroups establish H&S-related norms which they abide by over time. This may depend on a shared perception of an attached reward or consequence to the behaviour. Kines et al. (2011) mentioned that workers may express their commitment first to their workgroup before the organisation, hence making the workgroup the first point of socialisation for new workers and a strong influence on their H&S outcomes. Commitment in an organisation has been shown to provide an explanation of workers' H&S-related behaviour (Zin & Ismail, 2012). Co-workers' commitment to a shared goal has also been shown to enhance job satisfaction, emotional support and communication, as well as reducing role conflict; thus, it requires the support of management on all levels (Attiq et al., 2017).

Zohar and Luria (2005) highlighted the difference between compliance-based and commitment-based H&S management, noting that compliance-based H&S revolves around management ensuring that workers adhere to the H&S regulations set by the organisation or regulatory bodies. Managers often adopt compliance-based management system, owing to its lower cost of implementation, ready availability and ease of adaptability, despite being a weaker option. Yet commitment-based H&S complements compliance-based H&S, in the sense that it is closely related to OCB. In a high-risk work environment, OCB may include helping others and engaging in self-learning activities (Griffin & Hu, 2013). Because commitment-based H&S is intrinsic, it provides a better reliability in the prediction of safety outcomes when compared to compliance-based H&S. Emuze et al. (2016) add that compliance-based safety in developing countries must be backed up by H&S-related behaviour. As suggested by Lingard et al. (2012), workers' safety compliance is a stronger predictor of H&S performance when assessed at an organisational level because workers are very likely to interact more among themselves than with their supervisors. When workers collectively share the same perception regarding a prevailing safety climate, it is passed on as an acceptable organisational value, norm and behaviour in the workplace and may influence safety performance (Lingard et al., 2012). Positive work behaviour is often associated with OCB while the opposite is regarded as CWB. A healthy safety climate will enable OCB among workers, while a work environment with low levels of safety priority will continue to make it easy for CWB to fester among workers. CWB displayed by either individuals or, in some cases, groups in the workplace may put the organisation, individuals or workgroup in direct danger or harm and thwart the workgroup's ability to meet any set objective. The goal of any high-risk work environment is to build a safety climate which promotes OCB over CWB; this may be achieved by demonstrating safety priority among workers.

2.4.2 Organisational Citizenship Behaviour

Employees who are fully committed to the attainment of the organisation's safety goals and objectives may engage in behaviour that ensures OCB. The definition of OCB has, over time, witnessed some modification as it was initially seen as discretionary behaviour that, though not recognised or rewarded by the management, supports the effectual functioning of the organisation. As explained by Bolino and Klotz (2015), OCB extends beyond the assigned roles and sees workers voluntarily taking up additional job roles which ensure that the organisational objectives are achieved. Subsequently, it has been modified to mean behaviour which 'support[s] the social and psychological environment in which task performance takes place', given that studies have shown that employees are often at their best when they engage in OCB. According to Zabihi and Hashemzehi (2012), this behaviour is beneficial to the organisation and is often not imposed or elicited. Typical examples of OCB include peer group cooperation, the execution of extra roles or responsibilities without complaining and volunteering to help others. The concept of OCB has been researched in work environments such as business, academia and the police force. Findings suggest that high-risk work environments such as construction may exploit this concept in promoting a healthy safety priority among workers in the workplace.

2.4.3 Counterproductive Work Behaviour

Berry et al. (2012) referred to CWB as divergent behaviour willingly executed by workers that not only prevents the organisation from achieving its set goals and benefits but also puts the perpetrators of such acts and their colleagues in harm's way. As explained by Spector (2011), CWB can be defined from the employee's or organisation's perspective. From the organisational perspective, this is behaviour which inhibits the organisation from attaining its set H&S goal and objectives; yet when viewed from the employee perspective, the behaviour seeks to cause harm to both workers and the organisation.

The concepts of CWB and OCB are often regarded as negative and positive behaviours, respectively, displayed in the workplace. However, Bolino and Klotz (2015) argued that this view may be misleading and that workers are hardly either good or bad, neither is their behaviour. The same employee may display both OCB and CWB as influenced by prevailing work or personal circumstances. Though this argument is directed at the worker, it does not consider the fact that workers who engage in CWB undermine the attainment of workplace goals and objectives and such behaviour is explicitly categorised as negative irrespective of the circumstances which lead to its execution. Research has shown that though both constructs have a significant impact on the organisation, the existing relationship between both is negatively correlated (Ng et al., 2016). CWB is easily identified by the intended target; for example, behaviour such as attending to personal matters during work hours, taking additional break time and deliberately working slowly may be seen as behaviour that is targeted at the organisation. However, behaviour such as withholding vital information from the workgroup and spreading false rumours may be regarded as acts that are targeted at individuals.

2.5 Management Safety Support

The lack of management support towards workers' H&S has been reported as a cause of accidents and poor safety compliance on construction worksites (Awwad et al., 2016). Employees who enjoy the support of their organisation and management on H&S-related concerns have been found to have fewer accidents (Zin & Ismail, 2012). Management can provide the necessary support to employees on accident-prone worksites such as construction by demonstrating strong leadership skills on different levels, given that it has a direct and indirect relationship with unsafe work behaviour (Okorie, 2014). Top management demonstrating visible commitment to safety compliance may reduce workers' unsafe behaviour (Bosak et al., 2013).

Sadly, what is predominant in construction workplaces is a low level of management support, especially as it concerns workers. Workers who do not enjoy the support of management regarding their H&S may have low levels of safety priority and this may also contribute to enabling a workplace which increases the likelihood of hazards and accident occurrence. Owing to poverty levels in SSA, most workers are satisfied to simply earn the day's wage and do not demand a workplace which considers their H&S as a priority. They may also be unaware of what they must insist on as an acceptable work environment which supports their H&S. Another factor which is closely related is the poor employment rates in most countries in SSA. Frontline workers are often at the mercy of the contractor or supervisor who is also out to maximise profit since most contractors consider compliance to H&S regulations a waste of resources. Low levels of management safety support in work environments such as construction may reduce job satisfaction levels and should be considered as a contravention of workers' rights. Management may provide support to the workers in various ways which include but are not limited to the following:

- Provision of PPE
- H&S training and retraining

- Visible safety participation from top- and mid-level management
- Ensuring the establishment of a safety climate which promotes H&S-related behaviour

2.5.1 Safety-Related Behaviour

Safety-related behaviour has been described by Seo et al. (2015) as individuals' actions towards protecting themselves and others by adhering to H&S regulations. For example, workers who ensure that they wear the appropriate and recommended safety apparel at all times while at work may be said to engage in safety-related behaviour. An earlier submission by Kapp (2012) added that workers who volunteer for plant safety committees and actively participate by suggesting ways to improve the overall safety of the workplace make significant contributions to the organisation's H&S-related behaviour.

In a brief history of the evolution of H&S-related behaviour, Cooper (2009) stated that the application of the concept which began in the mid-1970s has witnessed a series of evolution. At first it was a top–down approach which saw the supervisor giving both positive and negative feedback on the workers' safety-related behaviour. Though this was a cheaper and easier method, it was quickly criticised for legitimate reasons attributed to workers' behaviour when the supervisors were not in sight. In the 1980s, a follow-up to this saw the creation of workgroups among employees who developed the entire process and conducted peer-to-peer observation with feedback. This again was criticised on the basis of the exclusion of the management process, thereby creating the perception that safety-related behaviour was an employee-based concept. The 1990s saw a marriage of both approaches in an employee–manager partnership where workers had the responsibility of monitoring their workgroups while supervisors monitored safety leadership behaviour, and both receive appropriate feedback. Despite the 1990s' birth of the employee–manager partnership of H&S-related behaviour, recent years have seen some construction managers still adopting one of the three methods mentioned earlier.

The concept of safety-related behaviour is a people-focused approach aimed at improving work-based safety compliance and reducing the likelihood of occurrence of fatalities in the workplace (Ismail et al., 2012a). Unsafe work behaviour stems from CWB which Berry et al. (2012) referred to as divergent behaviour willingly executed by workers, which not only prevents the organisation from achieving its set goals and benefits but also puts the perpetrators of such acts and their colleagues in harm's way. Okorie et al. (2016) argued that a revision from the top–down (management to the employee) approach to the enforcement of H&S regulation to a bottom–up (workers–management) approach will provide professionals with a better understanding of the reasons informing the decisions of workers to engage in CWB.

Dailey (2016) established that individuals may exhibit varying attitudes or behaviour while executing their jobs, depending on their perception of the degree of control that they feel they have over the situation. For instance, when certain individuals are faced with dull or meaningless work, they are most likely to engage in irrelevant behaviour, also known as immature disruptiveness or self-simulation which may undermine the productivity of other workers. In construction, workers with such personalities may engage in unsafe work behaviour that may not only put them and other workers' H&S at risk but also jeopardise the success of the project in its entirety.

To ensure effective safety-related behaviour which projects high safety priority levels in any workplace, factors such as leadership, monitoring and evaluation or feedback must be accorded equal and high priority. Jin and Chen (2013) mentioned that management commitment and the lack of sustained enthusiasm for the concept militated against the achievement of an efficient safety-related behavioural approach in the workplace. This implies that integrating safety-related behaviour from early on in the planning and design phases of the project may help the workers view the concept as part of their work (Hosseinian & Torghabeh, 2012). Sustainable safety-related behaviour can only be maintained when directed upwards; thus, the leadership style adopted in the workplace plays a significant role in creating awareness among workers of the benefits of safety-related behaviour and may influence their degree of acceptability of the concept (Okorie et al., 2016). Some employees, however, may not perceive the concept as important in the workplace and workers may also claim that they are ignorant of the same or simply believe that they cannot be involved in an accident during the course of their work (Zin & Ismail, 2012).

Frontline workers strongly believe that their role in H&S compliance is the sole responsibility of management and that they (workers) only engage in safety-related behaviour to avoid being fined or losing their jobs, not because they perceive it as their playing a role in their personal H&S (Choudhry & Fang, 2008). This perception must be corrected among workers as safety-related behaviour between management and workers displays a high level of safety priority in the organisation which, if sustained, will often result in improved general safety performance in the workplace. Choudhry (2014) contended that a safety-related behaviour initiative is one of the best approaches that can improve safety performance in high-risk work environments such as the construction industry and will enable workers to take responsibility for their personal H&S.

Further proactive steps in terms of the implementation of safety-related behaviour include baseline observations, safety training, follow-up observations, feedback and reinforcement. Management's safety priority in the workplace may impact workers' non-compliance behaviour (Zin & Ismail, 2012). Management is also often held responsible for the poor implementation and compliance to H&S regulations in the workplace. This emphasises the role management must play in visible safety prioritisation as an approach to improving safety compliance among workers. To ensure that workers are also held accountable for their safety-related behaviour, management may adopt approaches which ensure that frontline workers work safely. In their report, Cooper and Phillips (2004) suggested the use of a behavioural observation checklist (Table 2.1), where workers assessed each other by workgroups and tick whether what they saw was safe, unsafe or not seen.

Table 2.1 Behavioural observation checklist.

S/No	Behavioural observation	Safe	Unsafe	Not seen
1	Goggles must be worn when using nail gun			
2	Gloves must be worn when handling pallets and cases			
3	Electric knife must be used to cut film off the drums whenever possible			
	A manual knife may be used to start or finish the cut			
4	Nail guns only to be used in safe area behind protective barriers			
5	Loose strapping on pallets must be removed			
6	Protruding nails in pallets must be bent over safely or removed			
7	No jewellery should be worn (rings, watches, necklaces)			
8	Safety shoes must be worn			
9	Drivers of small forklift trucks must always remain at the front			
10	Reach trucks must not be driven with forks raised			
11	Horns must be sounded on blind corners			

Source: Adopted from Cooper and Phillips (2004).

This checklist resulted in an improved safety performance in the workplaces under study and may be further modified to address the requirements of individual workplaces or projects.

2.6 Safety Priority Issues in Developing Countries

Research findings from countries such as Nigeria and South Africa reveal that management perceives that they prioritise safety by providing PPE and safety guidelines; ensuring compliance; leading by example; crafting an organisational mission (safety) statement; constantly reminding workers to work safely; providing safety-related signage on project sites and implementing a practical, collaborative and stringent H&S culture among the workers. Therefore, it would be safe to assume that the accident rates should not be as high as they currently are in these countries; unfortunately, this is not the case. This picture may not be too different when compared with other countries in SSA. Management safety commitment and priority are an indication of safety-related behaviour on all levels, and in the mentioned countries, these levels appear low and require improvement.

Management often expresses a commendable understanding of safety commitment and priority; however, in practice, this may be lacking (Biggs et al., 2013; Windapo, 2013; Ismail et al., 2012b). The visible commitment of management to safety has the strongest influence on safety-related outcomes. When workers see that management actively participates in ensuring a safer workplace, they are motivated to display the same visible safety behaviour. Workers who have a positive perception of management's commitment to their H&S will also engage in safety-related OCB (Pinion et al., 2017). Management often alludes to poor safety-related behaviour such as not wearing the appropriate

PPE (Kheni & Braimah, 2014). Workers' safety commitment, priority and risk non-acceptance in the workplace may suggest that the behavioural norms that co-workers share in the workplace influence their safety behaviour and indicate the presence of a strong relational cohesion in the workgroup. Management must set up modalities and policies to ensure that this social exchange is maintained and explored in the interests of safety compliance and generally improved safety performance. Poor levels of relational cohesion in workgroups may be evidenced by workers' refusing to participate in housekeeping and failing to tackle and address the risk which may arise at work.

2.7 Conclusion

This chapter explored the safety priority from both management and workgroup perspectives. It established that workers' safety-related behaviour mirrors management safety behaviour. The chapter foregrounded the notion that safety matters are not operational alone. That is why the focus on strategic prioritisation of health, safety and well-being is advocated in practice and in this chapter. Safety prioritisation at the strategic level by management starts with having the right policies in place so that efforts required to engender a safe workplace where best practices are on display for everyone to emulate. Prioritisation of safety also includes appropriate designation of individuals on site in relation to safety roles, the training of all frontline workers and adequate use of the hierarchy of controls to deal with hazards and risks in operations. The discourse in the chapter, especially in relation to the safety priority of workers, management safety support and empirical reports, puts forward a working hypothesis. The working hypothesis that may be used to guide a future study is as follows: 'When workers perceive that management safety behaviour levels are low, they may engage in CWB; conversely, when they perceived that management upholds their H&S as a high and sustained priority, they may emulate such a perception with an improved level of safety compliance which eventually results in an improved safety performance for the organisation'. Management thus has the capacity to ensure that safety is prioritised in the construction workplace.

References

Ali NM, Jangga R, Ismai M, et al. (2015) Influence of leadership styles in creating quality work culture. *Procedia Economics and Finance* 31(15). Elsevier BV: 161–169. http://doi.org/10.1016/S2212-5671(15)01143-0

Alruqi WM, Hallowell MR and Techera U (2018) Safety climate dimensions and their relationship to construction safety performance: A meta-analytic review. *Safety Science* 109(June). Elsevier: 165–173. http://doi.org/10.1016/j.ssci.2018.05.019

Attiq S, Wahid S, Javaid N, et al. (2017) The impact of employees' core self-evaluation personality trait, management support, co-worker support on job satisfaction, and innovative work behaviour. *Pakistan Journal of Psychological Research* 32(1). National Institute of Psychology: 247–271.

Awwad R, El Souki O and Jabbour M (2016) Construction safety practices and challenges in a Middle Eastern developing country. *Safety Science* 83. Elsevier: 1–11. http://doi.org/10.1016/j.ssci.2015.10.016

Berry CM, Carpenter NC and Barratt CL (2012) Do other-reports of counterproductive work behavior provide an incremental contribution over self-reports? A meta-analytic comparison. *Journal of Applied Psychology* 97(3). American Psychological Association: 613–636. http://doi.org/10.1037/a0026739

Biggs SE, Banks TD, Davey JD, et al. (2013) Safety leaders' perceptions of safety culture in a large Australasian construction organisation. *Safety Science* 52. Elsevier: 3–12. http://doi.org/10.1016/j.ssci.2012.04.012

Bolino MC and Klotz AC (2015) The paradox of the unethical organizational citizen: The link between organizational citizenship behavior and unethical behavior at work. *Current Opinion in Psychology* 6. Elsevier: 45–49. http://doi.org/10.1016/j.copsyc.2015.03.026

Bosak J, Coetsee WJ and Cullinane SJ (2013) Safety climate dimensions as predictors for risk behavior. *Accident Analysis and Prevention* 55. Elsevier: 256–264. http://doi.org/10.1016/j.aap.2013.02.022

Cheng CW, Leu S-S, Lin CC, et al. (2010) Characteristic analysis of occupational accidents at small construction enterprises. *Safety Science* 48. Elsevier: 698–707. http://doi.org/10.1016/j.ssci.2010.02.001

Choudhry RM (2014) Behavior-based safety on construction sites: A case study. *Accident Analysis and Prevention* 70. Elsevier: 14–23. http://doi.org/10.1016/j.aap.2014.03.007

Choudhry RM and Fang D (2008) Why operatives engage in unsafe work behavior: Investigating factors on construction sites. *Safety Science* 46(4). Elsevier: 566–584. http://doi.org/10.1016/j.ssci.2007.06.027

Cooper MD (2009) Behavioral safety interventions: A review of process design factors. *Professional Safety* 54(2). American Society of Safety Professionals: 36–45.

Cooper MD and Phillips RA (2004) Exploratory analysis of the safety climate and safety behavior relationship. *Journal of Safety Research* 35(5). Elsevier: 497–512. http://doi.org/10.1016/j.jsr.2004.08.004

Dailey R (2016) The basics of organisational behavior and its relation to management. In: *Organisational behaviour*, pp. 1–44. Available from: www.ncbi.nlm.nih.gov/pubmed/2616327

Delegach M, Kark R, Katz-Navon T, et al. (2017) A focus on commitment: The roles of transformational and transactional leadership and self-regulatory focus in fostering organizational and safety commitment. *European Journal of Work and Organizational Psychology* 26(5). Routledge: 724–740. http://doi.org/10.1080/1359432X.2017.1345884

Emuze F, Linake M and Seboka L (2016) Construction work and the housekeeping. In: *Proceedings of the 32nd annual ARCOM conference* 1(September). Association of Researchers in Construction Management, pp. 537–546.

Griffin MA and Hu X (2013) How leaders differentially motivate safety compliance and safety participation: The role of monitoring, inspiring, and learning. *Safety Science* 60. Elsevier: 196–202. http://doi.org/10.1016/j.ssci.2013.07.019

Griffith A and Howarth T (2014) The nature of construction health and safety. In: *Construction health and safety management*. New York: Routledge, pp. 1–10.

Hosseinian SS and Torghabeh ZJ (2012) Major theories of construction accident causation models: A literature review. *International Journal of Advances in Engineering and Technology* 4(2). IJAET: 53–66.

Ismail F, Ahmad N, Afida N, Janipha NAI and Ismail R (2012a) Assessing the behavioural factors of safety culture for the Malaysian construction companies. *Procedia – Social*

and Behavioral Sciences 36(2012). Elsevier: 573–582. http://doi.org/10.1016/j.sbspro.
2012.03.063

Ismail F, Hashim AE, Zuriea W, Ismail W, Kamarudin H and Baharom ZA (2012b) Behaviour based approach for quality and safety environment improvement: Malaysian experience in the oil and gas industry. *Procedia – Social and Behavioral Sciences* 35(2012).
Elsevier BV: 586–594. http://doi.org/10.1016/j.sbspro.2012.02.125

Jiang L, Yu G, Li Y, et al. (2010) Perceived colleagues' safety knowledge/behavior and safety performance: Safety climate as a moderator in a multilevel study. *Accident Analysis and Prevention* 42(5). Elsevier: 1468–1476. http://doi.org/10.1016/j.aap.2009.08.017

Jin R and Chen Q (2013) Safety culture: Effects of environment, behavior & person. *Professional Safety* (May). American Society of Safety Professionals: 60–70.

Kapp EA (2012) The influence of supervisor leadership practices and perceived group safety climate on employee safety performance. *Safety Science* 50(4). Elsevier: 1119–1124. http://doi.org/10.1016/j.ssci.2011.11.011

Kheni NA and Braimah C (2014) Institutional and regulatory frameworks for health and safety administration: Study of the construction industry of Ghana. *International Refereed Journal of Engineering and Science* 3(2). IRJES: 24–34.

Kines P, Lappalainen J, Mikkelsen KL, et al. (2011) Nordic Safety Climate Questionnaire (NOSACQ-50): A new tool for diagnosing occupational safety climate. *International Journal of Industrial Ergonomics* 41(6). Elsevier: 634–646. http://doi.org/10.1016/j.ergon.2011.08.004

Lestari F, Sunindijo RY, Loosemore M, et al. (2020) A safety climate framework for improving health and safety in the Indonesian construction industry. *International Journal of Environmental Research and Public Health* 17(20). MDPI: 1–20. http://doi.org/10.3390/ijerph17207462

Liang H and Zhang S (2019) Impact of supervisors' safety violations on an individual worker within a construction crew. *Safety Science* 120. Elsevier: 679–697. http://doi.org/10.1016/j.ssci.2019.08.014

Lingard H, Cooke T and Blismas N (2012) Do perceptions of supervisors' safety responses mediate the relationship between perceptions of the organizational safety climate and incident rates in the construction supply chain? *Journal of Construction Engineering and Management* 132(2). ASCE: 234–241.

Mansur NNA, Ahmad MA, Ishaq HM, et al. (2011) Personality and organisational outcomes (organizational culture as a moderator). *International Journal of Academic Research* 3(6). IJAR: 54–60.

Mohammadi A, Tavakolan M and Khosravi Y (2018) Factors influencing safety performance on construction projects: A review. *Safety Science* 109(November 2018). Elsevier BV: 382–397. http://doi.org/10.1016/j.ssci.2018.06.017

Newaz MT, Davis P, Jefferies M, et al. (2019) The psychological contract: A missing link between safety climate and safety behaviour on construction sites. *Safety Science* 112.
Elsevier BV: 9–17. http://doi.org/10.1016/j.ssci.2018.10.002

Ng TWH, Lam SSK and Feldman DC (2016) Organizational citizenship behavior and counterproductive work behavior: Do males and females differ? *Journal of Vocational Behavior* 93. Elsevier: 11–32. http://doi.org/10.1016/j.jvb.2015.12.005

Okorie V (2014) *Behaviour-based health and safety management in construction: A leadership-focused approach.* PhD thesis, Nelson Mandela Metropolitan University, Port Elizabeth, South Africa.

Okorie V, Emuze F and Smallwood J (2016) Exploring the impact of team members' behaviours on accident causation within construction projects. In: *5th construction management conference, Nov. 2016*, Port Elizabeth, pp. 54–62.

Pinion C, Brewer S, Douphrate D, et al. (2017) The impact of job control on employee perception of management commitment to safety. *Safety Science* 93. Elsevier: 70–75. http://doi.org/10.1016/j.ssci.2016.11.015

Schein EH (2015) *Organizational culture and leadership.* 3rd ed. San Francisco, CA: Jossey-Bass.

Schwatka NV and Rosecrance JC (2016) Safety climate and safety behaviors in the construction industry: The importance of co-workers commitment to safety. *Work* 54(2). NLM/PubMed: 401–413. http://doi.org/10.3233/WOR-162341

Seo HC, Lee YS, Kim JJ, et al. (2015) Analyzing safety behaviors of temporary construction workers using structural equation modeling. *Safety Science* 77. Elsevier: 160–168. http://doi.org/10.1016/j.ssci.2015.03.010

Spector PE (2011) The relationship of personality to counterproductive work behavior (CWB): An integration of perspectives. *Human Resource Management Review* 21(4). Elsevier: 342–352. http://doi.org/10.1016/j.hrmr.2010.10.002

Strauch B (2015) Can we examine safety culture in accident investigations, or should we? *Safety Science* 77. Elsevier: 102–111. http://doi.org/10.1016/j.ssci.2015.03.020

Windapo A (2013) Relationship between degree of risk, cost and level of compliance to occupational health and safety regulations in construction. *Australasian Journal of Construction Economics and Building* 13(2). UTS ePress: 67–82. http://doi.org/10.5130/ajceb.v13i2.3270

Wu TC, Lin CH and Shiau SY (2010) Predicting safety culture: The roles of employer, operations manager and safety professional. *Journal of Safety Research* 41(5). Elsevier: 423–431. http://doi.org/10.1016/j.jsr.2010.06.006

Zabihi M and Hashemzehi R (2012) The relationship between leadership styles and organizational citizenship behavior. *African Journal of Business Management* 6(9). Academic Journals: 3310–3319. http://doi.org/10.5897/AJBM11.2809

Zin SM and Ismail F (2012) Employers' behavioural safety compliance factors toward occupational, safety and health improvement in the construction industry. *Procedia – Social and Behavioral Sciences* 36. Elsevier BV: 742–751. http://doi.org/10.1016/j.sbspro.2012.03.081

Zohar D and Luria G (2005) A multilevel model of safety climate: Cross-level relationships between organization and group-level climates. *Journal of Applied Psychology* 90(4). American Psychological Association: 616–628. http://doi.org/10.1037/0021-9010.90.4.616

3   Safety Empowerment

3.1 Introduction

Management empowers its employees in H&S when they show that they trust the employees enough to comply. There is a relationship between delegating authority to the employees in the form of leadership of subgroups and a reduction in accidents (Kines et al., 2011). The delegation of authority to the workers increases their sense of control over their work environment, which can further promote an enriched decision-making process where management considers the opinions of its workers in policy development (Karimi, 2011). As stated by Lawani et al. (2018), management must be seen to rise above the minimum legal requirement in H&S and aim for best practices by exploring the advantages of worker empowerment. Safety empowerment enables safety voice and observations not only from workgroup leaders but also from peer to peer. This makes it easy for supervisors to advise and guide accordingly, thereby boosting trust between supervisors and workers. Kheni and Braimah (2014) revealed that workgroup leaders are rarely empowered. A strong organisational culture may influence H&S behaviour; management may explore this to motivate frontline workers in favour of safety compliance and improved safety performance. Safety empowerment gives the workers a sense of belonging, appreciation and trust, which motivates them to accept H&S advice. This chapter discusses safety empowerment as a concept and presents findings from research conducted in two countries (Nigeria and South Africa) in SSA. The chapter concludes by highlighting how safety empowerment improves safety performance, especially among frontline workers on accident-prone construction sites.

3.2 Worker Empowerment

The empowerment of workers in any work environment has yielded positive outcomes for not only the organisation but also the employee as an individual as well. By definition, empowerment is a step adopted by management to develop employees' ability to think, act and make decisions in favour of their job roles or description (Al-Dmour et al., 2018). In another description, worker empowerment is regarded as a culmination of activities which provide, develop and improve

DOI: 10.1201/9781003361640-3

competence, productivity, discipline, attitude and work ethics to match the skill and expertise required in a job role (Iskandar et al., 2021). In today's highly competitive work environment, management must devise means to ensure that their workforce meets with the complexities of their field or run the risk of either phasing out or running out of business. It is no longer enough for the employer to employ new staff; they must empower them to perform optimally in accordance with their vision for the business. Employees who are empowered tackle their jobs with confidence because they trust themselves enough to make decisions by aligning with what their supervisors or managers suggest. An empowered workforce is able to focus on other equally pressing needs of the organisation because the employees can be trusted to carry on as if they were strictly supervised. Workers enjoy a commensurate job satisfaction level, and the organisation enjoys an improved level of quality and overall image, given that empowerment adds professionalism to the workers.

In terms of the concept of employee empowerment, though beneficial to all involved, Kagicia (2022) argued that an empowered worker may become too confident and arrogant, thereby making them difficult to be managed by the employer. This fear often makes the process difficult to embark on by the employer. Another legitimate fear is employee-poaching by another organisation. An employer may hesitate to improve the quality of their workers because they are concerned that the workers may feel they have outgrown the organisation and want to leave in search of better opportunities elsewhere. Workers also have expressed their own fear of empowerment as they perceive that it increases their workload and saddles them with the consequences of their decisions; they are now faced with accounting for everything they do and are unable to transfer the blame to their supervisors. Management again is worried about change and the fact that giving workers that much power will mean that they will have to divulge some trade secrets to the worker whom they do not trust enough to manage such sensitive information. Worker empowerment implies that management may have to let go of some of the control that they perceive to have over their workers: many managers are not very receptive to this change.

Considering the advantages and disadvantages of worker empowerment and whether worker empowerment is a bane or boon, Hechanova-Alampay and Beehr (2001) opined that workers who are empowered adapt to their expected workloads over time after recording improved job role clarity. This implies that the benefit of worker empowerment by far outweighs the fears of both employers and employees, while the impact of low levels of empowerment is better avoided than experienced. It is pertinent to note as well that worker empowerment is not only for employees; managers who wish to stay competitive in today's work environment must also be prepared to learn, relearn and unlearn. Empowering the workforce implies that they are not punished for errors but commended for trying and encouraged to do better next time. This does not imply that the worker will not be held accountable for the consequences of their actions but will enable them to control irresponsible behaviour at work. This eventually results in an organisation which is able to deal with market and business dynamics more efficiently.

3.3 Safety Empowerment in Accident-Prone Workplaces

Empowerment is hardly ever embarked upon to cater for safety in the workplace; rather, it is seen as a means to an end on the grounds that workers who are empowered at work are easily won over to working safely (Hechanova-Alampay & Beehr, 2001). In a high-risk work environment such as construction, there are hardly any standardised work processes, and it may be difficult to adopt approaches to safety such as is in other industries. Issues which border on workers' H&S are paramount in the construction industry and it is the responsibility of management to devise creative means to ensure that the workers comply with H&S regulations. Therefore, safety empowerment is an equally important approach to H&S compliance in the construction industry. Unfortunately, evidence in literature has shown that studies of safety empowerment emanating from construction still need to be improved. It has become evident that while gleaning from social sciences on the subject, authors must be wary of overemphasising related concepts without marrying them to safety climate.

Safety empowerment in construction may be described as the process of improving workers' capacity to comply with H&S regulations in the workplace. It is important that management communicate to the workers the role that they play in the organisation's safety performance to give them some sense of responsibility. Kagicia (2022) categorised the following types of empowerments for the purpose of safety:

- **Psychological empowerment**. This is people's perception of the level of value that they have to offer in an organisation. It also relates to the individual's perception of control they have over their work. Construction managers may devise means to boost workers' self-esteem by enabling the development of a work environment which promotes mutual respect and commends H&S-related behaviour. Psychological empowerment has been seen to be closely linked to individual job satisfaction.
- **Structural empowerment**. This refers to setting up organisational structures and opportunities which the employee may explore for the purpose of professional growth. Examples of such structures and opportunities include access to information, an efficient two-way feedback mechanism and access to equipment and tools when necessary.
- **Relational empowerment**. Strong social exchanges between management and workers result in the creation of a feedback system where both parties are able to speak up frankly concerning their jobs or assigned tasks. When this is achieved, workers are able to come forward with the challenges they face or errors they may have made without the fear of being punished and, in some cases, may be ready to account for the consequences of their decisions.
- **Team empowerment**. Workers in an organisation are the most important resource. Organisations normally depend on teams to execute certain tasks which may be cumbersome or difficult to achieve by an individual. An empowered team is able to rally around the leader to break down and offer solutions to complex tasks to ensure that the bigger objective of the organisation is achieved.

3.4 Safety Education

One of the most valuable methods of safety empowerment in a construction industry is through continuous safety learning and education for both management and workers. It is also found to predict safe work behaviour and participation which are both mediators in a safety climate (Jiang et al., 2010). H&S knowledge is often discussed along with H&S motivation, which is described as the construction workers' inclination towards performing their jobs safely. H&S learning results in improved H&S knowledge. According to Vinodkumar and Bhasi (2010), the workers' perception of management's support towards H&S-related activities does not necessarily predict H&S knowledge and motivation. Nevertheless, management must be conscious of the fact that factors such as the educational (formal) background of the workers may influence their grasp of H&S learning programmes which mostly focus on instructor-centric pedagogy and may yield little success in older workers (Loosemore & Malouf, 2019). Xu et al. (2019) added that similar factors such as work type, complexity, experience, age, motivation, emotion and learning strategy may impede the workers' learning abilities, yet the application of the acquired knowledge may result in an improved adherence to H&S regulations and safety-related behaviour. The adoption of learner-centric andragogy for the improvement and engagement of workers in H&S-related issues is highly encouraged as a form of safety empowerment.

H&S-related training in work environments such as construction may differ from the conventional system adopted in other workplaces. According to Griffin and Hu (2013), it is the extent to which supervisors motivate and promote H&S-related learning in an accident-prone workplace. H&S learning involves taking the necessary action in the event of near misses, incidents and from observing co-workers. Learning, which is also training, has been shown to improve skill competence and invariably boosts workers' self-esteem in their job roles by the adoption of specific methods such as approaching workers during work tasks and using scenarios, illustrations and teaching aids such as videos, photos, documents and equipment. This motivates the workers to open up to H&S conversations as opposed to keeping the workers in a formal environment with writing equipment and PowerPoint slide presentations (Zwetsloot et al., 2017). Demirkesen and Arditi (2015) maintained that organisations may adopt various approaches in H&S learning, but their goal should be to equip the worker with appropriate skills and knowledge to work safely. An accident and near misses reporting culture is one that motivates workers to discuss these occurrences freely and voluntarily without the fear of prejudice or blame (Zwetsloot et al., 2017).

3.5 Safety Learning

H&S-related learning in work environments such as construction may differ from the conventional system adopted in other workplaces. The use of such an approach in safety-related learning is not recommended for frontline construction workers because they may find it boring or too complex to understand owing to their perceived low levels of formal education. This may, however, be adopted for management and supervisors as the construction activity is one which requires training and retraining as the demand arises.

Demirkesen and Arditi (2015) added that organisations may adopt different approaches in achieving H&S learning; nevertheless, the goal should be to equip the workers with appropriate skills and knowledge to enable them to work safely. A reporting culture regarding reporting accident and near misses is one that motivates workers to discuss these occurrences freely and voluntarily without fear of prejudice or blame (Zwetsloot et al., 2017). This provides the management with data they can use in making H&S-related decisions. According to Lingard et al. (2010), though an increase in recorded near misses and accidents in the workplace may indicate a deteriorating safety performance, it also signals an improvement in the organisational safety incident reporting data. Incident reporting encourages workers to be vocal about accidents and near misses and also serves as a learning curve to the management; therefore, reporting should not be regarded in a negative light (Buck, 2011).

3.6 Safety Knowledge

The extent to which workers in the construction workplace understand the intricacies of H&S practices and procedures is referred to as H&S knowledge (Shen et al., 2017). When workers have a high level of safety knowledge, it becomes easy for them to engage in H&S-related behaviour which is a mediator to safety climate (Jiang et al., 2010). Safety knowledge is often discussed along with safety motivation, which is described as the construction workers' inclination towards performing their jobs safely. Safety learning is a strong predictor of an improved safety knowledge; according to Vinodkumar and Bhasi (2010), the workers' perception of management's support towards H&S-related activities does not necessarily predict H&S knowledge or motivation. Yet Loosemore and Malouf (2019) mentioned that factors such as the formal educational background of the workers may influence their grasp of H&S learning programmes which focus on *instructor-centric pedagogy* and may yield little success in older workers, implying that age is a factor which must be considered in safety learning and knowledge among construction workers. Xu et al. (2019) added that similar factors such as work type, complexity, experience, motivation, emotion stability and learning strategy, among others, may impede the workers' learning abilities. Nevertheless, the application of acquired knowledge may result in an improved adherence to H&S regulations in the workplace. The adoption of *learner-centric andragogy* for the improvement and engagement of workers in H&S-related issues is highly encouraged in such scenarios. A learner-centric andragogy is an approach towards organisational learning which captures the unique characteristics of the learner and uses these to design a knowledge transfer approach which facilitates learning and ensures maximum understanding. This is highly recommended for workplaces such as construction which deal with manual workers who are low on formal education levels.

3.7 Benefits of Safety Empowerment

Construction workers who are empowered in terms of safety become more valuable to the construction industry at large. This is because every construction team will stand to benefit from their level of empowerment. Considering that most frontline construction workers are hardly employed for long tenures, when they move to another project or another supervisor, they do so with the safety empowerment from the previous project. They bring this with them to the new project, thereby making it easier for the new supervisor to influence them towards H&S compliance.

1. **Delegation of authority**

Delegation of authority is the transfer of authority from a superior to an employee. It is an important part of the organisational day-to-day running, given that managers or supervisors may be unable to do tasks such as the effective monitoring of workers themselves (Kagicia, 2022). When management involves the leaders of workgroups in planning, execution and monitoring of H&S-related facets in the workplace, they can develop a realistic picture of what the workers' expect from them. This makes the workers more confident in their own ability regarding work safety. However, management must take certain key decisions themselves while supervising the workgroups (Jitwasinkul et al., 2016). This implies that H&S norms will be established by the workers over time, which will ensure H&S outcomes and reduce the likelihood of accident occurrence. Safety empowerment must not be viewed as a tool for controlling workers but as the workers' controlling their own work environment and accomplishing safety-related self-determination (Jitwasinkul & Hadikusumo, 2011). Workers' perception of control as delegated by management improves their confidence and enhances their ability to make positive H&S decisions, thereby reducing unsafe behaviour (Cavazza & Serpe, 2009). Workers who are able to contribute to organisational decision-making feel empowered and experience a sense of belonging. These types of workers are very likely to be loyal to the organisation and may engage in OCB or H&S-related behaviour.

2. **Workers' cooperation across hierarchical levels**

The construction process and its activities are a mix of various parties ranging from client and management to supervisors, subcontractors, workgroup leaders and workers. The success of the project will naturally depend on the cooperation enjoyed by all stakeholders (Fang & Wu, 2013). The project team will eventually break up into workgroups and these various workgroups will clearly state what goal is a priority for them to accomplish. For the attainment of group goals, co-workers may have to depend on themselves while different workgroups may depend on other groups for the collective attainment of the overall project objective. More importantly, management at different levels may utilise this strategy to ensure that H&S goals are not only captured in the subgroup objectives but also achieved. Zohar and Luria (2010) mentioned that workgroup leaders provide the link between management and the workers; therefore, it is important that there is a proven relationship among top and mid-level management and the group leaders where H&S commitment is a priority. According to Zohar and Luria (2003), the hierarchical nature of the workplace allows for H&S intervention from management at different levels. This implies that management must consider the dynamics of the workplace and set up interventions that encourage mutual cooperation.

3.   Mutual trust

Empowering workers fosters mutual trust and cooperation across the organisational hierarchy, that is between worker and worker and between management and workers. According to Parker et al. (2001), workers can make decisions such as timing and find the most effective method to adopt in the execution of their duties. However, the outcomes of management empowerment will largely depend on existing work conditions. Zohar and Luria (2010) mentioned that workers will build their H&S climate perceptions according to the emphasis that the organisation places on H&S.

Workers under a prevailing H&S climate will naturally create their own work-related cultural values in the organisation. The influence they have among themselves motivates them towards practising group H&S-compliance behaviour which results in improved H&S performance. Ultimately, this facilitates the creation of mutual trust among workers (Misnan & Mohammed, 2007). Trustworthiness among workers encourages teamwork as members of the workgroup influence each other's performance. Though there is evidence in literature that construction accidents are caused by the unsafe behaviour of workers, Mitropoulos and Memarian (2012) argued that accidents are also a factor of poor social cohesion among workers. The size of the workgroup may also influence mutual trust and the development of a sense of belongingness among workers in the workplace (Haldorai et al., 2020).

4.   Safety voice and error reporting

An organisation which has a healthy level of safety empowerment will be seen to display commendable levels of safety voice and error reporting. This is the case because workers are more confident in their job roles and are able to express themselves with regard to their individual and collective H&S. According to Lingard et al. (2010), an increase in recorded near misses and accidents in the workplace may indicate a deteriorating H&S performance, while at the same time signalling an improvement in the organisation's H&S reporting culture. The H&S reporting culture is an indicator of the H&S culture in an organisation. It encourages workers to be vocal about accidents and near misses and also serves as a learning curve for management. This factor should not be regarded in a negative light (Buck, 2011).

The importance of error reporting in construction is emphasised by De Silva and Rathnayake (2018), who noted that organisations with a poor safety climate and culture will often fail to report near misses and accidents that occur in the workplace. Accident reporting provides an important database for the organisation to monitor its safety performance and to provide compensation to affected workers. It also avails management of data pertaining to aspects of the workplace that need to be improved in favour of workers' H&S. Workers are often saddled with the responsibility of reporting accidents and near misses on their jobs. Unfortunately, research has shown that the number of accidents reported by the workers is never the actual representation of accident and near-miss statistics. Probst and Estrada (2010) mentioned that workers fail to report errors for reasons such as the fear of job loss, reprisals, as well as personal beliefs regarding accidents and injuries. This implies that the workers' perception of an unfavourable consequence may influence their decision to report an accident or near miss; for every reported accident, two are not reported. This leaves an organisation low on safety empowerment which, if left unchecked, will continue to allow conditions which increase the likelihood of accidents and hazards occurring.

Employee safety voice, according to Griffin (2016), refers to workers' ability to be vocal about their H&S-related concerns and give suggestions to management for improvement. Tucker et al. (2008) described the concept as communication intended at bringing change to an unsafe work environment. This is achievable through formal or informal channels directed towards management, co-workers, union or government officials. A worker's safety voice involves raising H&S-related concerns with management, addressing the safety committee or reporting unsafe work conditions to an OHS officer. Zhang et al. (2019) pointed out that workers may give up their H&S voice owing to the uncertainty of securing another job and may also leave if they perceive that their H&S voice is not heard or if the labour supply is insufficient. When workers perceive that management attaches high levels of importance to their safety voice, their job satisfaction levels and sense of belonging in the organisation increase, thereby positively influencing H&S compliance and OCB.

3.8 Improved Safety Performance

The primary objective of improved H&S performance in the workplace is to ensure that people are protected. As indicated by Cooper (2015), improved safety performance is likely if the benefit of H&S outweighs H&S compliance, implying that management must employ various methods to inspire their workers in favour of optimum H&S performance. H&S performance measures how a project performs against set H&S goals, which eventually results in H&S behavioural change in the workplace (Barbaranelli et al., 2015). According to Mohammadi et al. (2018), factors such as motivation, rules and regulations, H&S competency, H&S investment and cost, productivity, resources and equipment, work pressure, work conditions, climate and culture, attitude and behaviour, lessons learnt from past accidents, organisational factors and H&S programmes and management system may influence H&S performance.

Grill and Nielsen (2019) indicated that long-term planning is necessary for the attainment of improved H&S performance on construction worksites. It is important that workgroup leaders are included in the planning phase from inception to enable them to make recommendations with respect to safer work methods, given that they will be more conversant with what works for them in terms of their H&S. Safety leadership plays an instrumental role in safety performance of workers; management must be seen to be dynamic in its deployment of leadership styles, given that workers look up to them for inspiration. The introduction of industry-specific methods of learning and knowledge transfer in safety-related concepts is another method the management may deploy in the improvement of safety performance in the workplace. To improve safety empowerment in favour of safety performance, Lawani et al. (2018) proposed the following four factors:

- **Knowing**. This involves the worker acknowledging for themselves the value of their work, its requirement and standards. It is also required that the workers align this to what they perceive as important and their personal beliefs.
- **Doing**. This involves the workers' belief in their individual capabilities as they take on the job role.
- **Decision-making**. This concerns the workers' intentionally taking up the job role and accepting stipulated work methods, pace.
- **Influencing**. This is the workers' perception that they can influence certain outcomes within their work environment.

This summarily implies that workers who are allowed to have a say in their safety and work methods are not only better positioned to comply willingly with the set guidelines but are likely to experience enhanced job satisfaction as well. From management behaviour to safety justice, Table 3.1 shows nine factors that affect the safety performance of an organisation.

3.9 Safety Empowerment in Developing Countries

In developing countries in SSA, management seldom asks employees for their opinions before making decisions regarding H&S. They often adopt statutory H&S policies as guidelines for implementation, and there are rarely job- or task-specific H&S guidelines (Kheni & Bramah, 2014). This further implies that an organisational safety policy may only be available in theory but not in practical form. Where workers are involved, this is done during toolbox meetings. Management may perceive that they empower workers through giving them some level of authority on the worksite by communicating that they trust their judgement and ability. Findings in the current report reveal that safety empowerment levels in countries such as Nigeria and South Africa are low and require improvement. This was measured using the management safety empowerment dimension on the NOSACQ-50 safety climate assessment tool. The observed low empowerment of

Table 3.1 Factors influencing an improved safety performance.

Factor	Description
Top management's visible H&S behaviour	Management behaviour communicates the level of priority or importance workers attach to H&S in the workplace, thereby informing their perceptions and behaviours relative to the subject. When top management claims that workers and workplace H&S are at the top of its priorities, it is important that they follow their claim with actions for their workers to emulate. For example, it is insufficient for management to have it written as a policy that the wearing of PPEs by workers is mandatory and then proceed to dress inappropriately to work.
Safety leadership	Owing to the dynamics of the construction process, it is safe to say that a standardised approach to H&S leadership in construction may not be appropriate. It is paramount that management studies the type of work activity and the nature of its workers and apply an appropriate leadership style to suit the work environment. Leadership styles may be learnt by leaders as a form of personal development to be applied in the workplace for the purpose of improved H&S performance.
Safety justice	When workers perceive that they have been treated unfairly by management's decisions, they may compare the outcome with their perceived contribution or effort and may adjust their behaviour according to their perceptions. H&S justice ensures that workers are treated fairly and in a manner which encourages them to engage in safety-related behaviour such as error reporting and H&S communication to ensure improved safety performance.
Safety communication	The workplace must be set up to allow workers to communicate without fear of intimidation when they perceive that the work environment, equipment or conditions do not guarantee their H&S. This can be achieved by a set of relational exchanges, which in turn displays mutual respect between workers and management or supervisors, irrespective of their job role. Mutual respect often appeals to an innate human need, which may result in citizenship behaviour. Management may explore this in favour of improved H&S performance in the workplace.
Safety compliance behaviour	Workers must adhere to safety regulations and procedures by carrying out their work in a manner which displays that their H&S is a priority to themselves. H&S-compliant behaviour lies in activities in which workers engage to ensure H&S in the workplace.
Safety knowledge	Workers' understanding of H&S should inform or motivate H&S-compliant behaviour. Workers who are ignorant of the steps they need to take to ensure their H&S will either increase accident rates or engage in behaviour which puts themselves and their co-workers at risk of fatalities' occurrence.
Safety training	This is an H&S knowledge acquisition process which must be continually sustained in the workplace. H&S training is not a one-off activity; therefore, workers must be trained and retrained to maintain H&S awareness and to disseminate new safety knowledge.
Safety knowledge transfer	This involves the use of methods which suit the worker category and demography in the process of H&S education and training. The selection of this pedagogy must consider factors such as the workers' language, age, level of formal education, background, exposure, previous experiences, mental health and assimilation levels. However, these factors must not become hindrances to the transfer of H&S knowledge or used as a tool for work-related discrimination, considering that the construction industry provides a means of livelihood for a vast majority of persons to whom these factors may apply.
Safety communication	This refers to a set of relational exchanges which centre on H&S. It involves the process of sharing H&S knowledge among workers, which aids the safe performance of their work. H&S communication can be done either formally or informally to ensure workers' assimilation.

workers may further imply low levels of incident reporting, safety knowledge and safety compliance. The attending consequence of the observation is that workers may engage in unsafe behaviour, which increases the likelihood of incidents and accidents.

The decentralisation and equal distribution of authority in the workplace among workgroup leaders, however, increase workers' confidence in their ability to remain safe rather than management's attempting to control or force workers into H&S compliance. Empowered workers are also able to speak up about any H&S-related concerns they observe and even stop work if they confirm that their current work conditions places their H&S at risk (Jitwasinkul & Hadikusumo, 2011). Generally, it was observed that management has not created a work environment that supports H&S compliance in the provision of appropriate tools and equipment which support H&S performance. There is therefore a major opportunity for management to empower its workers regarding safety.

3.10 Conclusion

This chapter explored safety empowerment and how it can be improved for the benefit of construction workers. Given that developing countries share similar challenges and features regarding construction H&S compliance and enforcement, it can be concluded that safety empowerment levels in SSA are low and need to be improved. Different methods may be adopted for this purpose; however, most importantly, management must employ an approach which yields maximum results using appropriate safety education and knowledge transfer methods. Formal methods of learning may be suitable for top- and mid-level construction workers but may yield low result levels among frontline construction workers with low levels of formal education. Considering that frontline construction workers form a significant proportion of the industry, it is recommended that management consider this group a priority in interventions which seek to promote safety compliance. When management in an organisation fails to train frontline construction workers in the hope that they will be trained by another organisation, untrained workers will continue to move between organisations.

References

Al-Dmour R, Yassine O and Masa'deh R (2018) A review of literature on the associations among employee empowerment, work engagement and employee performance. *Modern Applied Science* 12(11). Canadian Center of Science and Education: 313–329. http://doi.org/10.5539/mas.v12n11p313

Barbaranelli C, Petitta L and Probst TM (2015) Does safety climate predict safety performance in Italy and the USA? Cross-cultural validation of a theoretical model of safety climate. *Accident Analysis and Prevention* 77. Elsevier: 35–44. http://doi.org/10.1016/j.aap.2015.01.012

Buck MA (2011) *Proactive personality and Big Five Traits in supervisors and workgroup members: Effects on safety climate and safety motivation.* Portland, OR: Portland State University. http://doi.org/10.15760/etd.268

Cavazza N and Serpe A (2009) Effects of safety climate on safety norm violations: Exploring the mediating role of attitudinal ambivalence toward personal protective equipment. *Journal of Safety Research* 40(4). Elsevier: 277–283. http://doi.org/10.1016/j.jsr.2009.06.002

Cooper D (2015) Effective safety leadership: Understanding types and styles that improve safety performance. *Professional Safety* (February). American Society of Safety Professionals: 49–53.

Demirkesen S and Arditi D (2015) Construction safety personnel's perceptions of safety training practices. *International Journal of Project Management* 33(5). Elsevier. APM and IPMA: 1160–1169. http://doi.org/10.1016/j.ijproman.2015.01.007

De Silva N, Rathnayake U and Kulasekera KMUB (2018) Under-reporting of construction accidents in Sri Lanka. *Journal of Engineering, Design and Technology* 16(6). Emerald Group Publishing: 850–868. http://doi.org/10.1108/JEDT-07-2017-0069

Fang D and Wu H (2013) Development of a safety culture interaction (SCI) model for construction projects. *Safety Science* 57. Elsevier: 138–149. http://doi.org/10.1016/j.ssci.2013.02.003

Griffin MA (2016) Safety climate in organizations: New challenges and frontiers for theory, research and practice. *Annual Review of Organisational Psychology and Organisational Behavior* 3(3). Annual Reviews: 197–212.

Griffin MA and Hu X (2013) How leaders differentially motivate safety compliance and safety participation: The role of monitoring, inspiring, and learning. *Safety Science* 60. Elsevier: 196–202. http://doi.org/10.1016/j.ssci.2013.07.019

Grill M and Nielsen K (2019) Promoting and impeding safety – a qualitative study into direct and indirect safety leadership practices of constructions site managers. *Safety Science* 114(April 2019). Elsevier: 148–159. http://doi.org/10.1016/j.ssci.2019.01.008

Haldorai K, Kim WG, Chang H and Li J (2020) Workplace spirituality as a mediator between ethical climate and workplace deviant behavior. *International Journal of Hospitality Management* 86(April 2020). Elsevier: 102372. http://doi.org/10.1016/j.ijhm.2019.102372

Hechanova-Alampay R and Beehr TA (2001) Empowerment, span of control, and safety performance in work teams after workforce reduction. *Journal of Occupational Health Psychology* 6(4). American Psychological Association: 275–282. http://doi.org/10.1037/1076-8998.6.4.275

Iskandar A, Subagdja O and Mubarok Z (2021) Policies and implementation of worker empowerment in Chinese companies in Konawe District, Southeast Sulawesi Province. *Geojournal of Tourism and Geosites* 38(4). University of Oradea: 1017–1025. http://doi.org/10.30892/GTG.38405-739

Jiang L, Yu G, Li Y, et al. (2010) Perceived colleagues' safety knowledge/behavior and safety performance: Safety climate as a moderator in a multilevel study. *Accident Analysis and Prevention* 42(5). Elsevier: 1468–1476. http://doi.org/10.1016/j.aap.2009.08.017

Jitwasinkul B and Hadikusumo BHW (2011) Identification of important organisational factors influencing safety work behaviours in construction projects. *Journal of Civil Engineering and Management* 17(4). Taylor & Francis: 520–528. http://doi.org/10.3846/13923730.2011.604538

Jitwasinkul B, Hadikusumo BHW and Memon AQ (2016) A Bayesian Belief Network model of organizational factors for improving safe work behaviors in Thai construction industry. *Safety Science* 82. Elsevier: 264–273. http://doi.org/10.1016/j.ssci.2015.09.027

Kagicia CN (2022) *Employee empowerment and job performance in National Polytechnics in Kenya*. Kenya: Jomo Kenyatta University of Agriculture and Technology.

Karimi R (2011) Reduce job stress in organizations: Role of locus of control. *International Journal of Business and Social Science* 2(18). Centre for Promoting Ideas, USA: 232–236.

Kheni NA and Braimah C (2014) Institutional and regulatory frameworks for health and safety administration: Study of the construction industry of Ghana. *International Refereed Journal of Engineering and Science* 3(2). IRJES: 24–34.

Kines P, Lappalainen J, Mikkelsen KL, et al. (2011) Nordic Safety Climate Questionnaire (NOSACQ-50): A new tool for diagnosing occupational safety climate. *International Journal of Industrial Ergonomics* 41(6). Elsevier: 634–646. http://doi.org/10.1016/j.ergon.2011.08.004

Lawani K, Hare B and Cameron I (2018) Empowerment as a construct of worker engagement and wellbeing. In: *Proceedings of the joint CIB W099 and TG59 conference coping with the complexity of safety, health, and wellbeing in construction*, Salvador, Brazil, 1–3 August 2018, pp. 388–376. Available from: https://www.irbnet.de/daten/iconda/CIB_DC31547.pdf

Lingard H, Wakefield R and Cashin P (2010) The development and testing of a hierarchical measure of project OHS performance. *Engineering, Construction and Architectural Management Construction and Architectural Management* 18(1). Emerald Group: 30–49. http://doi.org/10.1108/09699981111098676

Loosemore M and Malouf N (2019) Safety training and positive safety attitude formation in the Australian construction industry. *Safety Science* 113(August 2018). Elsevier: 233–243. http://doi.org/10.1016/j.ssci.2018.11.029

Misnan MS and Mohammed AH (2007) Development of safety culture in the construction industry: A conceptual framework. In: Boyd D (Ed.), *Procs 23rd annual ARCOM conference*, Belfast, UK, Association of Researchers in Construction Management, 3–5 September 2007, pp. 13–22. Available from: https://www.arcom.ac.uk/-docs/proceedings/ar2007-0013-0022_Misnan_and_Mohammed.pdf

Mitropoulos P and Memarian B (2012) Team processes and safety of workers: Cognitive, affective, and behavioral processes of construction crews. *Journal of Construction Engineering and Management* 138(10). ASCE: 1181–1191. http://doi.org/10.1061/(ASCE)CO.1943-7862.0000527

Mohammadi A, Tavakolan M and Khosravi Y (2018) Factors influencing safety performance on construction projects: A review. *Safety Science* 109(November 2018). Elsevier BV: 382–397. http://doi.org/10.1016/j.ssci.2018.06.017

Parker SK, Axtell CM and Turner N (2001) Designing a safer workplace: Importance of job autonomy, communication quality, and supportive supervisors. *Journal of Occupational Health Psychology* 6(3). American Psychological Association: 211–228. http://doi.org/10.1037/1076-8998.6.3.211

Probst TM and Estrada AX (2010) Accident under-reporting among employees: Testing the moderating influence of psychological safety climate and supervisor enforcement of safety practices. *Accident Analysis and Prevention* 42(5). Elsevier: 1438–1444. http://doi.org/10.1016/j.aap.2009.06.027

Shen Y, Ju C, Koh TY, et al. (2017) The impact of transformational leadership on safety climate and individual safety behavior on construction sites. *International Journal of Environmental Research and Public Health* 14(1). MDPI AG: 1–17. http://doi.org/10.3390/ijerph14010045

Tucker S, Chmiel N, Turner N, et al. (2008) Perceived organizational support for safety and employee safety voice: The mediating role of coworker support for safety. *Journal of Occupational Health Psychology* 13(4). American Psychological Association: 319–330. http://doi.org/10.1037/1076-8998.13.4.319

Vinodkumar MN and Bhasi M (2010) Safety management practices and safety behaviour: Assessing the mediating role of safety knowledge and motivation. *Accident Analysis and Prevention* 42(6). Elsevier: 2082–2093. http://doi.org/10.1016/j.aap.2010.06.021

Xu S, Zhang M and Hou L (2019) Formulating a learner model for evaluating construction workers' learning ability during safety training. *Safety Science* 116(July 2019). Elsevier: 97–107. http://doi.org/10.1016/j.ssci.2019.03.002

Zhang J, Mei Q and Liu S (2019) Study of the influence of employee safety voice on workplace safety level of small- and medium-sized enterprises. *Nankai Business Review International* 10(1). Emerald Publishing: 67–90. http://doi.org/10.1108/NBRI-08-2017-0045

Zohar D and Luria G (2003) The use of supervisory practices as leverage to improve safety behavior: A cross-level intervention model. *Journal of Safety Research* 34(5). Elsevier: 567–577. http://doi.org/10.1016/j.jsr.2003.05.006

Zohar D and Luria G (2010) Group leaders as gatekeepers: Testing safety climate variations across levels of analysis. *Applied Psychology: An International Review* 59(4). Wiley-Blackwell: 647–673. http://doi.org/10.1111/j.1464-0597.2010.00421.x

Zwetsloot GIJM, Kines P, Ruotsala R, et al. (2017) The importance of commitment, communication, culture and learning for the implementation of the Zero Accident Vision in 27 companies in Europe. *Safety Science* 96. Elsevier: 22–32. http://doi.org/10.1016/j.ssci.2017.03.001

4 Safety Justice

4.1 Introduction

The perception of fairness among workers in their dealings with management may result in organisational benefits or CWB. Dekker (2012) suggested that a just culture in the workplace ensures that deviant behaviour is treated in a manner which displays fairness irrespective of the position of the worker in the organisation. According to Haldorai et al. (2020), research has proved the existence of a significant relationship between the workers' perception of fairness and both H&S outcomes and CWB. When workers are convinced that they have not been treated justly, they engage in behaviour that may sabotage the progress of the organisation, their co-workers or both. In contrast, workers will give their best in the form of OCB if they perceive that they have been treated justly and fairly by their employer. This increases the organisation's H&S performance. A just culture in a high-risk work environment such as construction ensures that in the event of any safety-related violations, workers are confident that measures are taken by management to resolve or manage such acts with accountability and fairness. A work environment which does not display a visible justice culture may impact workers' error reporting. This further affects the quality of organisational safety performance. It is equally important that workers are fully aware of what is regarded as acceptable or unacceptable behaviour in the workplace. Workers must also know the implications of their behaviour. This chapter presents the various steps to be taken in achieving a just culture in the workplace and the implications of poorly managed workers' safety-related behaviour on general safety performance, starting with organisation justice in the next section.

4.2 Organisational Justice

When subjects such as justice come up in workplaces, they may not be unrelated to the fact that there is a perception of injustice over a workplace occurrence. A quick search of the keyword 'justice' in various dictionaries suggests that it is a principle adopted in addressing conflicts with fairness and, according to parties involved, what they deserve. This means that for justice to be administered there has to be a contravention of what is acceptable for the normal running of events.

DOI: 10.1201/9781003361640-4

In organisations, this may imply that set rules and regulations are broken either against the organisation or against the workforce. The concept of organisational justice describes how workers determine whether they have been treated fairly in the organisation and how their perception of fairness influences other organisational outcomes such as OCB, job satisfaction, CWB and organisational trust and commitment (Deepak & Perwez, 2018; Shah et al., 2022).

The quality of the relationship between an employee and an employer in any organisation has been described as social capital (Mahajan & Benson, 2013). Social capital is described as the ability of different workgroups in an organisation to agree among themselves to work together in order to achieve organisational objectives. This implies that how workers are treated in the workplace may make or mar their productivity levels and impact the organisation accordingly. The various dimensions of organisational justice include the following:

- **Distributive justice**. This refers to how employees assess their compensation in relation to their output; for example, employees may perceive that the pay they receive is not relative to the amount of work they put into the organisation. If an employee, for example, perceives that without them, a particular facet of the organisation will fail to run optimally, they may request better remuneration. Management's response to this may determine the employees' perception of fairness which in turn influences their work-related attitudes and behaviour. It is important to note that compensation may not only be of monetary value to the worker; receiving commendation or validation on their job from management or supervisors may be all that it takes for workers to perceive that they have received distributive justice. Another aspect of distributive justice is seen in situations where the supervisor commends or recommends one worker for compensation and ignores others who perceive that they have equally earned the benefits at stake. If the organisation has stipulated that some work-related behaviour is to be acknowledged or compensated in favour of the worker, this automatically becomes a goal. When this is achieved, management is seen to fulfil its part.
- **Interpersonal justice**. As the name implies, interpersonal justice relates to workers' perception of mutual respect between them and their peers and their employers. It is their expectation of courtesy, dignity and respect in the administration of justice within the organisation. The outcome of the justice administration may not really matter to the workers if they perceive that they were treated with dignity, respect and courtesy in the process. Furthermore, this perception either makes it easier for the workers to bear the consequences of their decisions or actions or influences their decision to build sustained social interaction for the completion of tasks.
- **Procedural justice**. This relates to clarity in the process of administering justice in the organisation; it shows the 'how' and 'why' and whether it is commensurate with the transgression. For example, if a worker is involved in deviant behaviour, the response of management to such behaviour must align with the gravity of the offence. Workers must see that the process has

followed a measure of control. According to Deepak and Perwez (2018), procedural justice must show consistency, non-bias, accuracy, correctability, representativeness and ethicality. When the process is clear and workers fully understand it, it makes it easy for them to accept the consequences of their behaviour. Another perspective to procedural justice may be found in management decisions such as compensation allocation; for example, workers may be quick to compare their wages or benefits against what is obtainable in sister organisations. The provision of work tools and equipment or favourable work conditions may also influence workers' perception of procedural justice in the organisation.

4.3 Safety Justice in the Workplace

It is difficult to discuss safety justice without first discussing what a just culture means. The concept of a just culture has received commendable attention in work organisations such as healthcare, aviation and nuclear energy. This is not unrelated to the fact that accidents which happen in such industries result in catastrophic outcomes for all stakeholders. A just culture is one which addresses the need for management to demonstrate to workers that the administration of justice in the workplace conforms to the principles of both moral and ethical standards (Heraghty et al., 2020; Dekker & Breakey, 2016). A just culture reflects statutes which govern accountability and discipline relating to unsafe behaviour in the workplace (Weiner et al., 2008; Clarke, 2011; Ambrose et al., 2019). In a just culture, workers are assured that they will be treated fairly in the event they engaged in an unsafe act and are willing to self-report because a culture of trust has been established between themselves and management. Workers are also able to express their views in investigations because they are confident that the system will protect them (Dekker, 2007).

Safety justice in a risk-prone workplace is important because it makes it easy for the management to collect data from workers for the purpose of improved safety performance through incident reporting. When workers lose confidence in managements' administration of justice, workers will cover up for themselves and their peers, creating a toxic work environment of we-versus-them. A we-versus-them attitude festers when management fails to fulfil its expectations regarding safety justice. Workers who sense that management will easily blame them without taking responsibility for the role they played in the creation of an environment which allowed unsafe acts to fester will naturally defend themselves. An ideal work environment will be one which upholds the administration of safety justice over a blame approach. According to Kines et al. (2011), accountability and blame must be adequately managed in safety justice to avoid a situation where workers are compelled to choose safety compliance over problem-solving.

OCB and CWB have been clearly found to be related to workers' perception of a just culture. Compelling workers to engage in H&S-related behaviour is commendable; however, when they engage in behaviour such as refusing to report unsafe acts or incidents owing to the fear of being punished, it becomes tantamount to CWB. On the other hand, workers who look out not only for them-selves but for their peers as well by simply reminding a colleague to act safely are considered to have engaged in OCB. The goal of management is to have a workforce which is not afraid to demonstrate safety-related OCB owing to the trust they have built with management.

Dekker and Breakey (2016) elucidated on two approaches of a just culture, namely retributive and restorative justice. In the former, management seeks to find who went against the rules or regulations and how they can be punished or disciplined without necessarily considering the factors which influenced them to commit the wrong. In contrast, restorative justice suggests that persons respon-sible for deviance should be allowed to discuss how it impacts them and how they can recover or heal from the consequences. In work environments such as construction, it will imply that a restorative justice approach affords management an opportunity to look holistically at the factors responsible for violations of H&S regulations among workers and to ensure that all persons involved are persuaded never to allow a recurrence or to reduce the chances of a repetition to the barest minimum.

It is safe to assume that most work environments adopt a retributive justice approach. This, according to Heraghty et al. (2020), stems from the perception of the management that punishing guilty persons over addressing the conditions which led to unsafe acts is good for the worker and the system. Furthermore, they suggest that punishment and negative reinforcements make the workforce rule-oriented where workers are fixated on obeying the rules to avoid being pun-ished. In such instances, workers are unable to express themselves regarding safer methods they perceive can be adopted to accomplish their tasks. Second, punish-ment puts the organisation's production targets at risk. In this case, the workers feel betrayed by the management's action and approach their work in a manner which suggests that they are simply in a survival mode. They may lose enthusiasm for their work and may not hesitate to leave as soon as an opportunity presents itself. Punishments harm the individual and inhibit organisational learning in the sense that if the workers perceive that they were dealt with unfairly, there is no limit to how much it may impact on them mentally. Workers may become vis-ibly depressed and management, if keenly observant, will notice the new attitude towards their work. Other workers, fearing being treated in like manner, may fail to report incidents to management, thereby denying the organisation data or infor-mation which could be used for the purpose of organisational learning.

4.4 Incident Reporting

The purpose of incident reporting in accident-prone workplaces is organisational learning and improved safety performance. According to Frazier et al. (2013), when the act of incident reporting is volitional, it may indicate that the workers are taking personal responsibility not only for their own safety but also that of other colleagues. The aviation sector has successfully implemented a non-punitive approach to incident reporting. This may be related to the reduced numbers of recorded accidents in the field (Weiner et al., 2008). Another sector following closely on this is the health sector (Heraghty et al., 2020). The quality of data collected during incident reporting must come from employees who are confident in the system, given that they submit data freely without fear of being punished or disciplined. Shao et al. (2019) mentioned that the quality of data collected during incident reporting may be unreliable owing to factors such as a poor audit system. Mahajan (2010) indicated that acceptable incident report data for organisational learning should depend on the following activities:

- **Data input**. This must be independent and from a non-punitive approach.
- **The data**. The process of data collection must be structured to allow the reporters to express themselves freely. The use of close-ended questions is

Table 4.1 Factors inhibiting incident reporting.

Organisational factors	These include organisational values, beliefs and policies impacting incident reporting. They also include factors in the organisation which act as barriers to a healthy reporting culture. For example, a complex organisational structure or reporting line may make it difficult for workers to report an incident.
Work environment	Some features in the workplace may serve as a barrier to incident reporting. Examples include the use of heavy surveillance cameras, which may make workers feel like robots, or a shortage in the number of workers assigned to a task, among others.
Process and systems of reporting	A complex reporting system will deter workers from reporting incidents while a simple one will facilitate the process. This may or may not be influenced by the level of information required from the reporter and whether the submission process is manual (paper-bound) or electronic.
Team factors	Members of a team who understand the importance of incident reporting will encourage other members to engage. On the other hand, members who have suffered any form of victimisation from the process will inadvertently discourage others too. Healthy communication among team members may also facilitate the process.
Knowledge and skills	This refers to the workers' skills and training on the job task that enable them to know that an incident needs to be reported. Owing to low levels of safety knowledge, some workers may have encountered near misses several times without realising that these need to be reported.
Individual characteristics	Some personalities are more predisposed to expressing their views than others. Workers who are more open and have high self-esteem levels will be more inclined to report incidents than those who are withdrawn and non-confrontational.
Fear of adverse consequences	A retributive or restorative approach towards justice administration in the workplace will definitely determine how responsive workers are towards incident reporting.

Source: Adapted from Archer et al. (2017).

discouraged, whereas open-ended questions allow the submission of individual and personalised reports based on the workers' version of events.

- **Analysis.** A standardised methodology is highly recommended for the purpose of analysis. This enables the report to be easily turned into a lesson. Experts in the field of safety come together to interrogate the data by observing any links between the incident and the system. The sole aim of this phase is to generate meaningful learning outcomes.
- **Feedback.** This involves the use of multiple sources where all parties involved come together, lower their defences and drop the blame attitude to brainstorm on meaningful ways they can learn from the incidents and how to prevent a reoccurrence. The goal of the feedback phase is to derive positive outcomes from the incidents. This will motivate other workers to continue to engage in incident reporting.

Incident reporting does not necessarily translate into a measure of safety for any industry; rather, it provides valuable information that the organisation may use to improve its safety performance rating. An incident report cannot be used for comparing one organisation's safety record to that of another. In workplaces such as construction, no two projects are the same as each project always has its unique sets of attributes which must be properly managed (Mahmoudi et al., 2014). An attempt to compare two organisations based on their incident reports will be tantamount to comparing lemons to oranges; they might look alike but are completely different. Data collection and management in any endeavour may be too cumbersome or expensive if not managed properly. This is one factor that management must note in order to enhance the organisation's incident reporting system. A safety data collection team may be constituted to decide on what comprises an incident, whether it qualifies to be reported and how it must be reported (Haas & Yorio, 2019; Avram et al., 2015). This is to forestall any clogging of the system with irrelevant data which amounts to low-value reports.

4.5 Accountability in a Just Culture

Accountability in the workplace is a process whereby the individuals involved in an event are given an opportunity to give an explanation as to why they did what they did freely without fear. McCall and Pruchnicki (2017) described accountability as a process of being able to recount an event to someone else. In safety justice, accountability may be described as an ability to express an event of failure and being confident to accept the role played by the reporter without the fear of intimidation or punishment. Accountability may be organisational and individual. The former demands that the workplace is able to accept the role it played in the breakdown of events which led to incidents or the role it played in enabling the conditions which increased the likelihood of accidents occurring. Individual accountability, on the other hand, focuses on the worker. According to Wachter (2013), the goal must be to attain collective accountability, where everyone involved is able to admit to the role they played which resulted in the accident for learning purposes and a work environment which shuns a name-blame-shame approach.

Accountability in a just culture, according to Dekker (2007), may be forward-looking or backward-looking. In forward-looking accountability, the goal is to find opportunities for organisational learning and using the supposedly sad events to set up policies which ensure that the likelihood of reoccurrence of a similar or the same event is either eliminated or reduced to the barest minimum. A backward-looking accountability looks for guilty persons and blames or punishes them for the events that transpired. Accountability mostly surfaces in the events of failure. However, when failure is perceived as an opportunity for growth and learning, the events which led to failure become a springboard for success. McCall and Pruchnicki (2017) referred to the same as retrospective and prospective types of accountabilities, with retrospective accountability addressing past events and failing to identify and address latent failures in the system to prevent future mishaps. Although it offers an opportunity for the organisation to identify and categorise errors, it nevertheless leads to a blame and fear culture which may inhibit incident reporting. Prospective accountability is future seeking for the purpose of learning and planning within safety systems in a just culture. Both restorative and prospective accountability types are essential in setting up an efficient justice system, with restorative justice assessing multiple accounts to understand what must be done to restore any trust lost and attempting to fix any broken-down relationships between management and workers.

4.6 Safety Justice in Developing Countries

How management manages safety-related behaviour on construction sites was reported in the Nigerian and South African study. The results from the study demonstrate the approach adopted in dealing with such behaviour in the workplace. This relates to how management either rewards or punishes compliance or non-compliance to H&S standards respectively. This communicates a degree of fairness in management's dealings with workers. The results reveal that there is a low level of safety justice administration and this needs to be improved. Workers will not report incidents or near misses for fear of sanctions or victimisation. Others admit that management looks for guilty persons, not accident causes. This suggests that a retributive justice approach is practised in construction organisations in the study area. Zwetsloot et al. (2017) attested in their study on Zero Accident Vision that safety justice issues must be focused on sharing and learning. They contended that though companies perceived the importance of safety justice in the safety climate, very little has been done to identify a standardised method for its implementation. The reporting of accidents and near misses in accident-prone environments is deemed mandatory; however, workers' perception of management's actions may influence the reporting culture. De Silva et al. (2018) lamented in their study that owing to the complexities surrounding accident reporting it is often overlooked and leads to more occurrences This situation results in a lack of feedback that could be used by management in the development of better techniques and ideas to improve safety performance.

The results from the Nigerian and South African study suggest that managers in construction issue warnings, raise queries and impose penalties such as wage and salary cuts. In the worst cases, workers are removed from the site if deviant behaviour persists. This is a confirmation that the approach adopted by management in safety justice administration is more punitive than restorative. It further suggests reasons for the low levels or non-existence of incident reporting on most construction worksites.

4.7 Conclusion

Safety justice in a risk-prone work environment is vital to the safety climate, given that there will be errors, which may be human- or system-related. How management handles issues which stem from violations or deviant behaviour from workers will go a long way in either sustaining the organisation's safety performance or resulting in further blaming or shaming of either the system or the workers. The goal of every safety climate is to sustain a justice culture where workers understand the benefits of incident reporting and accountability that are aimed at organisational learning. Management must build workers' trust in the system to the level where they are able to come forward with their safety-related concerns without the fear of punishment or intimidation. A work environment which has healthy safety justice will result in more OCB and this is what every management should aspire to. The construction industry must define what safety justice means to it considering the unique characteristics inherent in the industry and then design a model which enables seamless implementation.

References

Ambrose ML, Rice DB and Mayer DM (2019) Justice climate and workgroup outcomes: The role of coworker fair behavior and workgroup structure. *Journal of Business Ethics* 172(1). Springer Netherlands: 79–99. http://doi.org/10.1007/s10551-019-04348-9

Archer S, Hull L, Soukup T, et al. (2017) Development of a theoretical framework of factors affecting patient safety incident reporting: A theoretical review of the literature. *BMJ Open* 1: 1–16. http://doi.org/10.1136/bmjopen-2017-017155

Avram E, Ionescu D and Mincu CL (2015) Perceived safety climate and organizational trust: The mediator role of job satisfaction. *Procedia – Social and Behavioral Sciences* 187. Elsevier BV: 679–684. http://doi.org/10.1016/j.sbspro.2015.03.126

Clarke DM (2011) Just culture: Balancing safety and accountability. *Journal of Risk Research* 14(1). Routledge: 143–144. http://doi.org/10.1080/13669877.2010.506272

Deepak PS and Perwez SK (2018) Measurement of organization justice scale and its dimensions on white collared professionals – an empirical analysis. *International Journal of Creative Research Thoughts* 6(1). IJCRT: 543–557.

Dekker SWA (2007) *Just culture: Balancing safety and accountability.* Farnham, UK: Ashgate.

Dekker SWA (2012) *Just culture: Balancing safety and accountability.* 2nd ed. Boca Raton, FL: CRC Press.

Dekker SWA and Breakey H (2016) 'Just culture': Improving safety by achieving substantive, procedural and restorative justice. *Safety Science* 85. Elsevier: 187–193. http://doi.org/10.1016/j.ssci.2016.01.018

De Silva N, Rathnayake U and Kulasekera KMUB (2018) Under-reporting of construction accidents in Sri Lanka. *Journal of Engineering, Design and Technology* 16(6). Emerald Group: 850–868. http://doi.org/10.1108/JEDT-07-2017-0069

Frazier CB, Ludwig TD, Whitaker B, et al. (2013) A hierarchical factor analysis of a safety culture survey. *Journal of Safety Research* 45. Elsevier: 15–28. http://doi.org/10.1016/j.jsr.2012.10.015

Haas EJ and Yorio PL (2019) The role of risk avoidance and locus of control in workers' near miss experiences: Implications for improving safety management systems. *Journal of Loss Prevention in the Process Industries* 59(March). Elsevier: 91–99. http://doi.org/10.1016/j.jlp.2019.03.005

Haldorai K, Kim WG, Chang H and Li J (2020) Workplace spirituality as a mediator between ethical climate and workplace deviant behavior. *International Journal of Hospitality Management* 86(April 2020). Elsevier: 102372. http://doi.org/10.1016/j.ijhm.2019.102372

Heraghty D, Rae AJ and Dekker SWA (2020) Managing accidents using retributive justice mechanisms: When the just culture policy gets done to you. *Safety Science* 126(February). Elsevier: 104677. http://doi.org/10.1016/j.ssci.2020.104677

Kines P, Lappalainen J, Mikkelsen KL, et al. (2011) Nordic Safety Climate Questionnaire (NOSACQ-50): A new tool for diagnosing occupational safety climate. *International Journal of Industrial Ergonomics* 41(6). Elsevier: 634–646. http://doi.org/10.1016/j.ergon.2011.08.004

Mahajan A and Benson P (2013) Organisational justice climate, social capital and firm performance. *Journal of Management Development* 32(7). Emerald Publishing: 721–736. http://doi.org/10.1108/JMD-12-2010-0091

Mahajan RP (2010) Critical incident reporting and learning. *British Journal of Anaesthesia* 105(1). Elsevier: 69–75. http://doi.org/10.1093/bja/aeq133

Mahmoudi S, Ghasemi F, Mohammadfam I, et al. (2014) Framework for continuous assessment and improvement of occupational health and safety issues in construction companies. *Safety and Health at Work* 5(3). Elsevier Science BV: 125–130. http://doi.org/10.1016/j.shaw.2014.05.005

McCall JR and Pruchnicki S (2017) Just culture: A case study of accountability relationship boundaries influence on safety in HIGH-consequence industries. *Safety Science* 94. Elsevier: 143–151. http://doi.org/10.1016/j.ssci.2017.01.008

Shah N, Anwar S and Irani Z (2022) The impact of organisational justice on ethical behaviour. *International Journal of Business Innovation and Research* 12(2). Inderscience: 240–258.

Shao B, Hu Z, Liu Q, et al. (2019) Fatal accident patterns of building construction activities in China. *Safety Science* 111. Elsevier: 253–263. http://doi.org/10.1016/J.SSCI.2018.07.019

Wachter RM (2013) Personal accountability in healthcare: Searching for the right balance. *BMJ Quality and Safety* 22(2). BMJ: 176–180. http://doi.org/10.1136/bmjqs-2012-001227

Weiner BJ, Hobgood C and Lewis MA (2008) The meaning of justice in safety incident reporting. *Social Science and Medicine* 66(2). Elsevier: 403–413. http://doi.org/10.1016/j.socscimed.2007.08.013

Zwetsloot GIJM, Kines P, Ruotsala R, et al. (2017) The importance of commitment, communication, culture and learning for the implementation of the Zero Accident Vision in 27 companies in Europe. *Safety Science* 96. Elsevier: 22–32. http://doi.org/10.1016/j.ssci.2017.03.001

5 Safety Leadership

5.1 Introduction

Over the years, the concept of leadership in the workplace has enjoyed attention from both scholars and professionals alike. The reasons are that the importance of strong leadership in a work environment dictates the pace at which organisational goals are met. It is commonplace to blame almost every dysfunction in an organisation on a weak leadership structure or presence. Leadership enables the execution of authority and ensures that things are done following the organisation's goals and objectives. When leadership fails, everything else follows (Martínez-Córcoles et al., 2011); it is for this reason that fields of study such as construction cannot afford to fail in the design of a leadership structure which is safety-focused on the workplace. Effective leadership demonstrates knowledge of the appropriate style to be used in motivating frontline workers towards safety compliance.

This chapter explores the various types of leadership approaches which can be modified by management and supervisors in pursuit of improved H&S compliance and general safety performance in the workplace. Construction-related studies, especially those relating to workers, must understand the attributes of human behaviour. Factors such as the personality of an individual are what make them unique; some people are more prone to risky or unsafe behaviour than others; leaders who understand the attributes of different personalities are able to devise appropriate approaches for spurring them towards safety compliance.

5.2 Safety Leadership in Construction

In accident-prone workplaces such as construction, the importance of strong safety leadership can never be overemphasised. Safety leadership may be viewed as a process whereby leaders in an organisation are able to influence the workers towards the attainment of H&S-related goals in the organisation (Oah et al., 2018). It is the main factor which sets up a team to succeed and influences OCB, which in turn determines safety value (Cooper, 2015a). Safety leadership involves directing workers in a manner which results in improved safety-related outcomes. Safety leadership seeks to motivate workers through a leader or an organisation in adopting various approaches to achieve workforce H&S compliance for the

DOI: 10.1201/9781003361640-5

benefit of overall improved safety performance. Safety leadership works in tandem with safety motivation. It is said to be positive where good safety-related behaviour is recognised and commended by management and negative where unsafe behaviour is ignored, reprimanded or criticised. Though there is a vast amount of literature relating to leadership in the workplace, there is still a dearth in industry-specific studies, especially regarding safety leadership among frontline construction workers (Barg et al., 2014).

The importance of human power or resources in any organisation cannot be overemphasised, given that organisations depend on people to achieve their objectives. It is therefore imperative that leaders understand the various types of people they lead as well as their characteristics so they can lead them towards achieving safety-related goals as deemed appropriate. Martínez-Córcoles et al. (2011) suggested that the use of traditional methods in leadership, especially in accident-prone workplaces, has outlived its usefulness and encouraged management to explore non-traditional methods in the attainment of organisational safety. To this it must added that the approaches adopted by management must be ethical and ensure that human rights and respect for construction workers are upheld to the highest standards. Understanding other disciplines which focus on human behaviour such as psychology and sociology will go a long way in helping management devise means of motivating workers in favour of improved safety performance. No two individuals are identical; people all have distinct personalities which may be influenced by different factors. Personalities have been studied by psychologist and a variety of traits have been grouped according to different personality types. These may be used to an extent to predict how an individual who has a dominant personality trait may behave.

Psychologists have different categorisation of personalities; literature abounds with descriptions of each and arguments for and against different facets of the typology. Exploring these arguments is beyond the scope of this book. However, the Big Five personality trait theory appears in a number of safety-related sources of literature. According to Almlund et al. (2011), the Big Five personality traits represent the longitude and latitude of personalities and highlight that 'personality traits are the relative enduring patterns of thoughts, feelings and behaviours that reflect the tendency to respond in certain ways under certain circumstances'. Following on this definition, it is evident that what works for individual A may not be the case for B. This may suggest that the work ahead for safety leadership is challenging but not impossible. Pourmazaherian et al. (2017) mentioned that while personalities differ significantly, an individual may display a range of traits along different personalities. Literature shows evidence that some human resources management practices require the matching of job descriptions to human personalities, while others have argued regarding the ethicality of this practice, given that potential workers may be discriminated against because of their personality. In safety leadership, what is important for leaders is to understand the various personality traits so that they are better-informed when they identify certain behaviours among their workers and can better position themselves to motivate these workers to H&S compliance.

5.2.1 Big Five Personality Traits

Safety leadership must understand the tendencies of different personalities, especially regarding unsafe behaviour in the workplace. For example, Kozako et al. (2013) reported on the Big Five personality traits, namely extraversion, agreeableness, conscientiousness, neuroticism and openness to experience, and how these personalities demonstrate CWB. The findings of this study are presented in Table 5.1.

The interesting study is by no means exhaustive and serves only as a guide to leaders regarding what to expect from human behaviour and how to inspire these personalities towards improved levels of H&S compliance. In an accident-prone work environment, leaders must be aware and ensure that they take up the mantle of inspiring workers in favour of improved safety performance in the workplace.

All humans have an innate desire to be treated with respect and dignity. Leaders who are able to tap into this need may achieve higher safety leadership outcomes over those who do not. Fernández-Muñiz et al. (2014) mentioned that workers

Table 5.1 How different personalities demonstrate CWB.

Big Five personality trait	CWB
Extraversion	This personality trait is categorised as positive. Individuals with high levels of extraversion are easily confident in themselves, display evidence of dominance, are active, excitement seeking and have low tendencies to display anger. In risk-prone workplaces, the extraversion personality trait is easily assumed as one who may not be involved in CWB.
Agreeableness	People who score high on this trait are regarded as compassionate and cooperative, displaying an inability to disagree. They are not inclined to be suspicious or antagonistic towards others. The relationship between this personality trait and CWB was found to be negative, implying that persons high on agreeableness may demonstrate low levels of CWB.
Conscientiousness	This personality trait shows high levels of self-discipline and may aim for achievement over expectations. This category was found to have a positive relationship with CWB, implying that individuals who are high on conscientiousness may be likely to demonstrate CWB.
Neuroticism	This trait relates to an individual's emotional stability. People who score high on this personality trait have demonstrated high levels of emotional exhaustion. The study found a positive relationship between this personality trait and CWB.
Openness to experience	People conforming to this trait are creative, imaginative, willing to explore new events and curious. This personality type also displays high levels of emotional exhaustion which may lead to CWB.

Source: Adapted from Kozako et al. (2013).

interpret open communication with their leaders as high-quality relationship and this easily results in outcomes such as an improved commitment towards H&S compliance among workers. Safety leadership may explore this relationship for the promotion of safety-related behaviour in the workplace. Emuze and Small-wood (2017) suggested treating people in construction with dignity, respect and justice, indicating that some leaders treat frontline construction workers, especially migrant workers, as tools and a means to an end. In safety climate, the workers' behaviour is hardly unrelated to their leaders' and treating people with respect must come from a genuine place of authentic care. Leaders who are indifferent about their workforce may see similar behaviour from workers; for example, workers may approach activities such as housekeeping with an attitude of indifference. This may increase hazards and injuries' occurrence on project sites.

As rightly captured by Wu et al. (2010), excellent safety leadership does not happen magically; it is a result of sustained effort and growth. Leaders must be open to training and retraining on how to motivate and inspire workers, particularly in terms of H&S-related objectives in the workplace. To improve safety performance in the workplace, the commitment of all project leaders to H&S issues must be evaluated. Managers on all levels must re-evaluate their leadership styles and ensure that their particular style accommodates H&S motivation for their workers (Okorie et al., 2016). The consistency and quality of top management leadership significantly influence mid-management's and employees' H&S-related behaviour (Zin & Ismail, 2012). Top management involvement in H&S-related behaviour communicates a high H&S priority to the worker and inspires H&S-related behaviour.

5.3 Transactional Safety Leadership

Leadership literature establishes types of leadership styles, one of which is the transactional leadership style. According to Kapp (2012), this is an approach based on Skinner's theory of operant conditioning whereby the leader sets work-related goals subject to either reward or penalty and then monitors the worker to ensure that the goal is achieved. The worker receives the reward if this is the case or is penalised in the event of any failure to deliver. Penalties may also be in the form of correctional feedback in an attempt for the leader to achieve improved behaviour. Transactional leadership is synonymous with motivating compliance through incentives and sanctions. This may be suitable in cultures with higher regard for relational boundaries. Transactional leaders enforce the powers vested in their authority to obtain results; employees or workers only follow because they have to (Cooper, 2015b).

The transactional leader is able to exert this approach effectively by adopting three behavioural dimensions: first, contingent reward, which in safety leadership refers the leader's ability to clearly spell out to the workers what they must do if they want to achieve the set reward. Second is management-by-exception: active. This represents the leader's ability to monitor workers' performance and note the corrections therein. Last, a transactional safety leader could practise management-by-exception: passive. This relates to the degree to which the leader looks out for non-compliance among workers in order to exert punitive action (Delegach et al., 2017).

Transactional safety leadership distinguishes between performance requirements and anticipated outcome. Such leaders perceive that all workers are responsible for the role assigned to them and adopt performance-monitoring approaches such as coaching, positive affirmations and the provision of any required support to reinforce the H&S-related behaviour (Cooper, 2015b). Over the course of time, a transactional safety leader gradually teaches the worker that H&S-related behaviour must be attached to a reward and the workers eventually perceive that they are only able to work safely if there is an attached incentive (Bello, 2012).

5.4 Transformational Safety Leadership

Transformational leadership, according to Deinert et al. (2015), is a meaningful and creative exchange between leaders and their followers which aims at instilling overall vision-driven work behaviour in the organisation. Transformational leaders focus on shaping and transforming an organisational culture on the basis that they know the goals they want to achieve and will employ all means possible to ensure that it is done (Cooper, 2015b). As opined by Martínez-Córcoles et al. (2011), transformational safety leadership is more relatable in safety climate studies. It involves a closer relational exchange between the leader and workers. This approach is mostly suitable in cultures where workers do not have a high regard for relational boundaries between management, supervisor and workers.

A transformational leader is able to achieve safety leadership through four main behavioural approaches, as explained by Delegach et al. (2017). First is their ability to exert idealised influence, which is the level to which they are able to present themselves as mentors to workers by demonstrating admirable safety-related behaviour. Second is their ability to become inspirational motivators, which in this regard is the degree to which they envision a desirable safety-related future and are able to communicate this to the workers in an inspiring manner. Transformational safety leaders may also be intellectually stimulating, which lies in their ability to question what is obtainable when they perceive it may not be to the advantage of the worker. They are also able to take risks and encourage the workers to work creatively and innovatively. Last, they have an individualised consideration for each worker. They attend to the developmental needs of the workers because they understand the workers' unique needs.

According to Niessen et al. (2017), transformational leadership involves the leaders first being what they want to see reflected in their workers. In workplaces such as construction, this implies that the transformational safety leader will first demonstrate a visible H&S-related behaviour by *walking the talk*. The initial idea influencing this style of leadership is that the leader appeals to the morality and values of the worker so that they transcend their personal interests for the collective good of their workgroup or organisation to achieve work-related goals (Bello, 2012). The argument for and against incorporating safety goals into the literature of transformational leadership has enjoyed attention among scholars. Mullen and Kevin Kelloway (2009), however, noted that including safety in transformational leadership literature has been observed to improve safety-related outcomes as a transformational leader may be passive regarding safety and active on other organisational goals. In construction, transformational leaders must ensure that safety-related outcomes remain a top priority.

Fernández-Muñiz et al. (2017) argued that the approach to transformational safety leadership which seeks to appeal to workers emotionally for the benefit of H&S-related behaviour may be insufficient to achieve the anticipated results. Leaders must actively participate in the change they desire to see in the workplace. Transformational safety leadership should be treated as a motivational approach to encourage worker participation in H&S-related behaviour. The goal of transformational safety leadership is to have a team of workers who do not need to be incentivised before they engage in H&S-related behaviour such as OCB. Over the course of time they understand the implication of complying with H&S regulations and will do so whether their supervisor or workgroup leader is present or not. A work environment which has attained this level of safety compliance will naturally transfer it to new members in the organisation because an organisational culture which values safety has been established.

5.5 Servant Safety Leadership

A servant leader builds personal relationships with the workers and ensures that workgroup members garner the support they need by ensuring that there is an effective safety communication in place that workers may maximise when their performance begins to fall short of what is expected. This approach unleashes workers' ability to make a difference in their H&S-related outcomes (Cooper, 2015b). This approach is worker-focused and not leader-focused because the leaders position themselves to serve the workers systematically by making available all they need to achieve H&S-related goals optimally. Servant safety leadership aims to set the workers up for success by meeting their H&S-related needs; for example, a servant safety leader ensures that the tools and equipment workers need to work safely are provided as and when needed. Workers in this kind of work environment are confident that they have the support of their leaders on H&S-related goals, further improving worker engagement and team performance. In a servant safety leadership structure, workers are seen to participate willingly as members of safety committees, participate in safety meetings such as toolbox meetings and seek improved ways to work safely by expressing their ideas to management on how they perceive a task can be accomplished more safely. Servant safety leaders consider the needs of workers over their own, serve charismatically, express an attractive vision and expect high levels of performance. The workers then in turn respond with trust, loyalty, confidence and respect; this is often displayed as OCB.

In assessing the various types of safety leadership, it may be out of place to perceive that one is better than the other. Rather, each style may have its time and

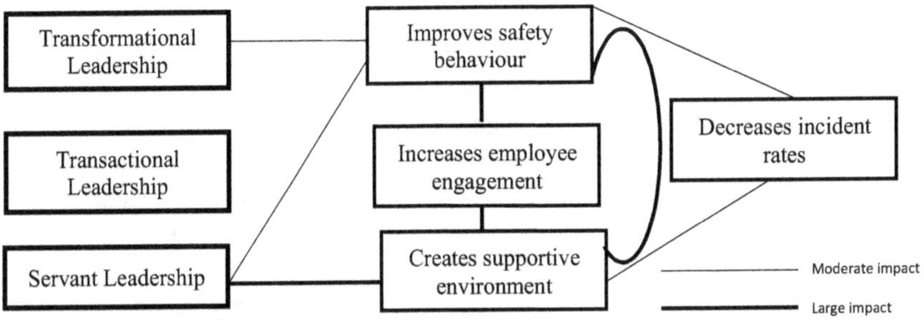

Figure 5.1 Positive impact of safety leadership styles.

Source: Adopted from Cooper (2015a).

place when and where it must be deployed (Cooper, 2015b). An effective leader is one who understands the human needs of the workforce and modifies or adopts an appropriate style to ensure that safety-related goals are achieved. Indeed, the most preferred or suitable leadership style will be a hybrid safety leadership approach in which the benefits of all the styles are maximised for organisational goals which, in a safety climate, will improve safety performance.

5.6 Workers' Safety Leadership Behaviour

One of the goals of safety leadership in high-risk work environments such as construction should focus on the decentralisation of leadership. Given that the construction activity requires the collective effort of different work subgroups, it is important that safety leadership is present at the workgroup level. Porter et al. (2016) discussed workers' motivation to lead in the work environment and concluded that the perceptions and beliefs of the worker may drive their motivation and willingness to take up leadership responsibilities at work. By implication, it can be deduced that the safety climate may predict workers' motivation to lead. Safety leadership has been shown to be closely related to safety performance; therefore, when workers perceive that the organisation has high levels of safety leadership, they will be confident to take up similar roles when appointed (Cooper, 2015a; Fernández-Muñiz et al., 2017).

Construction work environments may be designed so that the manual workers are able to rise through the ranks following a structure. This may also be constituted on a worker performance rating scale to motivate the workers to become more active in H&S compliance with the anticipation that they may be considered for a leadership position in the course of their jobs. However, this may only be practicable in work environments where manual workers have permanent tenure. Skeepers and Mbohwa (2015) argued that the role of safety leadership in the workplace cannot be passed down to workers on the grounds that top management are the main drivers of safety performance. The authors' standpoint is based on the decentralisation and delegation of authority down to workgroup levels. This does not exclude management or supervisors from the implementation of safety compliance in the workplace, but it reduces the line of command and makes it easier for management to enforce compliance using the workgroup leaders.

Workgroup leaders may require some form of training to assume the position of leadership at that level, given that they may require improved skills on people management. They may have a commendable perception of safety leadership in the workplace; nevertheless, they will need to become effective safety leaders of other members of the workgroup. Kapp (2012) highlighted that a supervisor's leadership style influences workers' perception of safety leadership. It then becomes imperative that supervisors can demonstrate a safety leadership style that projects H&S-related outcomes as a consistent priority in the workgroup. This further promotes a shared perception of safety among work group members and enhances H&S-related behaviour.

The advantage of having workgroup safety leadership in place is that the leader is relatable to other workgroup members, given that the leader is a peer. Group members freely discuss safety-related concerns with the confidence that it will reach management and be accorded the attention it deserves. Incident reporting becomes easier for workers and they are better positioned to advocate collectively for acceptable work conditions when they perceive that they are not being treated fairly by management. This further improves the safety voice of frontline workers. The line of reporting is shortened and enhances effective safety communication in the workplace. When factors such as these are established, workers become confident in their ability to engage in H&S-related behaviour and the organisation may record higher levels of safety performance. There may be a perceived notion among management that workers who share a collective sense of reasoning may decide to frustrate work-related goals if they are unable to have their way. However, management which has demonstrated an acceptable level of safety leadership need not be concerned because workers' attitude and behaviours reflect those of the management.

5.7 Conclusion

This chapter discussed safety leadership and linked the different styles of leadership approach to safety in construction and other risk-prone workplaces. The ability of workers to take up leadership roles in their workgroups was also discussed. Management of construction operations should explore how to improve incident reporting and the organisation's safety performance. Management must understand the attributes of different personalities in the workplace so they are better positioned to inspire them to safety leadership. There is no leadership approach that is superior to others; however, safety climate-related studies find transformational leadership to be more appropriate in accident-prone workplaces. It is recommended that safety leaders create a hybrid of approaches to suit the needs and uniqueness of individual projects and align these with the organisational culture.

References

Almlund M, Duckworth AL, Heckman JJ and Kautz TD (2011) Personality psychology and economics. In: Hanushek E, Machin S and Woessman L (Eds.), *Handbook of the economics of education*. Amsterdam: Elsevier, pp. 1–181.

Barg JE, Ruparathna R, Mendis D, et al. (2014) Motivating workers in construction. *Journal of Construction Engineering* 2014(12). Hindawi: 1–11. http://doi.org/10.1155/2014/703084

Bello SM (2012) Impact of ethical leadership on employee job performance. *International Journal of Business and Social Science* 3(11). Center for Promoting Ideas: 228–237.

Cooper D (2015a) Effective safety leadership: Understanding types and styles that improve safety performance. *Professional Safety* (February). American Society of Safety Professionals: 49–53.

Cooper D (2015b) Effective safety leadership. *Professional Safety* 60(2). American Society of Safety Professionals: 49–53.

Deinert A, Homan AC, Boer D, et al. (2015) Transformational leadership sub-dimensions and their link to leaders' personality and performance. *The Leadership Quarterly* 26(6). Elsevier: 1095–1120. http://doi.org/10.1016/j.leaqua.2015.08.001

Delegach M, Kark R, Katz-Navon T, et al. (2017) A focus on commitment: The roles of transformational and transactional leadership and self-regulatory focus in fostering organizational and safety commitment. *European Journal of Work and Organizational Psychology* 26(5). Routledge: 724–740. http://doi.org/10.1080/1359432X.2017.1345884

Emuze F and Smallwood J (Eds.). (2017) *Valuing people in construction*. London: Routledge. http://doi.org/10.4324/9781315459936

Fernández-Muñiz B, Montes-Peón JM and Vázquez-Ordás CJ (2014) Safety leadership, risk management and safety performance in Spanish firms. *Safety Science* 70. Elsevier: 295–307. http://doi.org/10.1016/j.ssci.2014.07.010

Fernández-Muñiz B, Montes-Peón JM and Vázquez-Ordás CJ (2017) The role of safety leadership and working conditions in safety performance in process industries. *Journal of Loss Prevention in the Process Industries* 50(October). Elsevier: 403–415. http://doi.org/10.1016/j.jlp.2017.11.001

Kapp EA (2012) The influence of supervisor leadership practices and perceived group safety climate on employee safety performance. *Safety Science* 50(4). Elsevier: 1119–1124. http://doi.org/10.1016/j.ssci.2011.11.011

Kozako IN, 'Ain MF, Safin SZ and Rahim ARA (2013) The relationship of Big Five personality traits on counterproductive work behaviour among hotel employees: An exploratory study. *Procedia Economics and Finance* 7(Icebr). Elsevier BV: 181–187. http://doi.org/10.1016/s2212-5671(13)00233-5

Martínez-Córcoles M, Gracia F, Tomás I, et al. (2011) Leadership and employees' perceived safety behaviours in a nuclear power plant: A structural equation model. *Safety Science* 49(8–9). Elsevier: 1118–1129. http://doi.org/10.1016/j.ssci.2011.03.002

Mullen JE and Kelloway K (2009) Safety leadership: A longitudinal study of the effects of transformational leadership on safety outcomes. *Journal of Occupational and Organizational Psychology* 82(2). Wiley-Blackwell: 253–272. http://doi.org/10.1348/096317908X325313

Niessen C, Mäder I, Stride C, et al. (2017) Thriving when exhausted: The role of perceived transformational leadership. *Journal of Vocational Behavior* 103. Elsevier: 41–51. http://doi.org/10.1016/j.jvb.2017.07.012

Oah S, Na R and Moon K (2018) The influence of safety climate, safety leadership, work-load, and accident experiences on risk perception: A study of Korean manufacturing workers. *Safety and Health at Work* 9(4). Elsevier: 427–433. http://doi.org/10.1016/j.shaw.2018.01.008

Okorie V, Emuze F and Smallwood J (2016) Exploring the impact of team members' behaviour on accident causation within construction projects. In: *Proceedings of the 5th construction management conference*, Port Elizabeth, South Africa, 28–29 November 2016, pp. 54–62.

Porter TH, Riesenmy KD and Fields D (2016) Work environment and employee motivation to lead. *American Journal of Business* 31(2). Emerald Publishing: 66–84. http://doi.org/10.1108/ajb-05-2015-0017

Pourmazaherian M, Baqutayan SMS, Idrus D, et al. (2017) The role of the big five personality factors on accident: A case of accidents in construction industries. *Journal of Science, Technology and Innovation Policy* 3(2). Penerbit UTM Press: 46–55. Available from: www.jostip.org/index.php/jostip/article/view/62

Skeepers NC and Mbohwa C (2015) A study on the leadership behaviour, safety leadership and safety performance in the construction industry in South Africa. *Procedia Manufacturing* 4(Iess). Elsevier BV: 10–16. http://doi.org/10.1016/j.promfg.2015.11.008

Wu TC, Lin CH and Shiau SY (2010) Predicting safety culture: The roles of employer, operations manager and safety professional. *Journal of Safety Research* 41(5). Elsevier: 423–431. http://doi.org/10.1016/j.jsr.2010.06.006

Zin SM and Ismail F (2012) Employers' behavioural safety compliance factors toward occupational, safety and health improvement in the construction industry. *Procedia – Social and Behavioral Sciences* 36. Elsevier BV: 742–751. http://doi.org/10.1016/j.sbspro.2012.03.081

6 Safety Commitment

6.1 Introduction

The concept of commitment has enjoyed scholarly attention from every field of endeavour. This is by no means surprising as commitment is vital to every organisation. Safety climate conversations regarding safety commitment are often hinged on safety leadership; this is not unrelated to the fact that effective safety leadership is incomplete without visible safety commitment. Commitment is a strong indicator of good leadership. In an organisational safety climate, commitment is what workers see and what informs their perception of management safety behaviour. Workers build their perception of safety from management's visible safety commitment. It is safe to say that safety commitment is a determining factor of safety climate. Workers who perceive that management is genuinely committed to their safety will respond by engaging in H&S behaviour; on the other hand, management which does not display a visible commitment to workers' safety will experience low levels of safety compliance from workers. Safety commitment is perceived by some schools of thought as the most critical construct of the safety climate. If management or supervisors only communicate their commitment to safety without attending behaviour, workers may interpret this as acceptable. This chapter focuses on management safety commitment and its implications for workers' behaviour. It further discusses how both management and workers may demonstrate visible safety commitment in the workplace to motivate workers in favour of H&S-related behaviour with an outcome targeted at an overall safety performance for the organisation.

6.2 Management Safety Commitment

In a work environment with high accident propensity, the demonstration of safety commitment is imperative. An organisation's safety commitment indicates how far it is willing to go or invest in the promotion of safety and accident-prevention strategies (Zwetsloot et al., 2017). Management safety commitment, supervisors' safety behaviour and workgroups' H&S-related behaviour are among the factors which Amponsah-Tawiah and Mensah (2016) mentioned as influencing safety commitment in the workplace. Globally, H&S in the workplace is a complex

DOI: 10.1201/9781003361640-6

issue, especially among top management as they must ensure it remains a top priority in an accident-prone organisation such as construction. The commitment of management to H&S-related issues in the workplace communicates to the workers the extent of management's concern for their personal and collective safety. Management must be seen to demonstrate positive safety attitudes because employees' perceptions of these may influence workers' safety participation and compliance levels. Workers' perception of management safety commitment has also been seen to result in reduced accident levels while organisational safety-related policies and practices positively impact on workers generally (Geldart et al., 2010).

The role of supervisors in demonstrating organisational safety commitment often comes up in related conversations. This is because in many organisations, supervisors represent management; depending on the size of the organisation, some workers may rarely interact with top management personnel (Hardison et al., 2014). To the workers, visible safety management is only displayed by the supervisor. The role of the supervisor in safety commitment is the key in the management of organisational H&S. According to Yule et al. (2007), workers may engage in H&S-related behaviour when they perceive that their supervisors are fair to them and also comply with organisational safety policies and regulations. Supervisors' safety behaviour demonstrates how far they are willing to go to ensure that unsafe acts among workers are kept to the barest minimum as well as their readiness to acknowledge and commend workers with high levels of H&S-compliant behaviour (Kapp, 2012). Huang et al. (2017), however, argued that supervisors' closeness to the workers in terms of line of command may influence their levels of safety compliance; for example, supervisor A's demonstration of positive safety behaviour may have no effect on workgroup B. Workers in group B may not observe high levels of safety compliance and participation if their own supervisor does not show visible safety commitment.

In coming up with measures to help leaders and supervisors perform optimally in accident-prone workplaces, Reb et al. (2014) mentioned the concept of mindfulness, which they described as a state of being fully aware of one's current state emotionally and psychologically while interacting with others in a state that suggests that they are not being judged. Increased mindfulness levels have been seen to result in the improved quality of social interaction. It follows that leaders or supervisors who are able to lead mindfully may demonstrate to their workers a degree of respect and dignity which workers may interpret as commitment (Emuze & Smallwood, 2017). In safety climate studies, leaders may focus on safety-related outcomes to communicate their commitment levels to workers in order to inspire H&S-related behaviour among workers and a generally improved safety performance level for the organisation. Hardison et al. (2014), however, suggested that supervisors may not have the required training on how to manage the workers under them. This may likely provide more explanations on why this level of leaders has constantly borne blame for workers' low levels of safety participation. In considering how important supervisors are to management, it is necessary that their leadership competence must be improved upon by management

Figure 6.1 Management safety commitment.

Source: Modified from Ismail et al. (2012).

to enable them to represent top management in the light in which they wish to be seen by the management. Management must also invest in a supervisor monitoring strategy by enabling feedback from workers on supervisors' H&S-related behaviour to keep certain excesses in check and to ensure that all workers on the project site have improved H&S compliance levels.

Wei et al. (2015), among other scholars who have shown commendable interest in worker commitment to safety, have highlighted the importance of management's focus on workers' H&S-related behaviour as it contains the answers to improved safety performance and compliance in the workplace. Pinion et al. (2017) added that workers' job control levels may further influence their perception of management safety commitment. Workers who are overwhelmed by their job roles may subsequently begin to consider engaging in CWB owing to low levels of job control. Workers with low levels of job control may also perceive that management is not committed to their personal safety and this perception may influence their H&S-related behaviour. It is important that in management's demonstration of safety commitment they are able to assign work-related tasks commensurate to the workers' ability and job roles to improve job control levels and motivate workers' safety commitment.

Management may demonstrate safety commitment in the workplace by means of the following:

- Demonstrating visible safety participation across all managerial levels.
- Supporting all H&S-related interventions, especially those which aim at improving safety compliance levels among workers; for example, safety training and acknowledging and rewarding safety-related behaviour, among others.
- Demonstrating their support for workers' safety by providing commensurate safety leadership and all resources required for H&S implementation.

- Behaving in a manner which demonstrates an unshakeable belief in the benefits of H&S compliance in the workplace.
- Ensuring that safety-related values, goals and resources in the workplace are adequately implemented and controlled.
- Establishing a positive attitude towards the H&S of workers as this demonstrates an equal priority and importance with regard to other project goals.

6.3 Safety Compliance and Participation

In the safety climate literature, safety compliance and participation are often mentioned as useful indicators of safety performance. He et al. (2019) and Griffin and Hu (2013) indicated that both concepts are closely related; however, safety compliance refers to *in-role safety-related behaviours which are voluntary* while safety participation relates to *out-role safety behaviours which are voluntary*. According to Kapp (2012), safety compliance refers to behaviour which indicates an adherence to safety regulations and procedures and is often at the core of maintaining safety-related outcomes in the workplace. Safety participation, on the other hand, may not necessarily contribute to workplace safety but aids the development of workplace H&S. According to Al-Bsheish et al. (2019), workers' perception of management safety commitment directly influences safety compliance and incident rates, further influencing safety performance as well. Typical safety performance in construction may include putting in efforts to ensure that the workplace safety is improved, helping co-workers and promoting a safety programme within the organisation.

Safety compliance may be viewed as a statutory obligation of the worker in an accident-prone workplace while safety participation may not necessarily be compulsory; for example, workers who refuse to engage in a safety campaign in the workplace may have demonstrated low levels of safety performance. However, this does not translate to their having low levels of safety compliance. Though safety compliance is strongly related to safety participation, it may not be an indicator of safety participation. In developing a measurement scale for safety performance, Dearmond et al. (2011) found no statistical difference between safety compliance correlations and that of safety participation, implying that the patterns of observed relationships between both variables are the same. Griffin and Neal (2000) also indicated that the main antecedents to safety performance were motivation, skill and knowledge. By this it can be postulated that a worker who has high levels of safety motivation, skill and knowledge may have to demonstrate positive levels of safety compliance and participation to yield an anticipated safety performance. Management must consider these antecedents in order to advance safety commitment in workers.

Griffin and Neal's antecedent to safety performance is safety competence, which is a general perception of an individual's skill, knowledge, qualification and training. Kvalheim and Dahl (2016) mentioned that there is a positive causal relationship between safety competence and safety compliance. This implies that management may boost safety compliance levels among workers by investing in their safety competence. It is also noteworthy that workers' perception of management safety competence may also influence their safety compliance and participation levels. Zohar (1980) indicated that the qualification of a safety officer may influence the safety-related outcomes among workers. When management or their representatives in the position of supervisors or foremen demonstrate low safety competence levels, safety compliance and participation levels will become low, given that they are unable to demonstrate an attending safety leadership and commitment in the workplace. In many countries, general construction workers have low levels of formal education. It is important that management takes note of the unique characteristics of their general workers in developing a safety training and education scheme or strategy. Previous findings have shown that management often adopts formal training methods which workers may find monotonous and boring, thereby resulting in low levels of understanding of what is being taught them. Adopting informal methods for safety training among general workers may yield an improvement in workers' safety knowledge and further safety compliance and participation.

6.4 Workers' Safety Commitment

Employees' commitment to an organisation has been described as their personal attachment to the organisation they work for and their willingness to remain and identify with the organisation (Amponsah-Tawiah & Mensah, 2016). This commitment is often informed by a perceived personal benefit or interest they stand to achieve or gain from the organisation which binds them to the employer for a period. An employer may influence the workers' commitment by giving them a sense of belonging, incentives or mutual respect. According to Meyer and Allen's (1991) three component commitment model, an employee may have affective commitment, which is emotionally driven; continuance commitment, which is described as an employee's inability to leave an organisation due to the perceived cost attached to such decision, and last, an employee may display normative commitment, which is the inability to leave an organisation due to the perception of obligation to the same. Workers who find organisational principles favourable are more likely to remain committed to what the organisation represents. It is therefore important to note that an organisation must offer some form of value to the worker to achieve their commitment to yielding the output the organisation anticipates (Cohen, 2007).

Lawani et al. (2018) described workers' safety commitment in an accident-prone workplace as workers' perception that management's safety behaviour aligns with their personal need for safety. These categories of workers will remain loyal to the organisation because they want to, and their decision to remain loyal is not influenced by incentives but by feelings of satisfaction that they are significant contributors to the organisation's safety performance. Such safety commitment influences OCB and other H&S-related behaviour among workers.

In the safety climate literature, safety commitment has been established to be dependent on management visible safety behaviour. However, it is important to note that co-workers may also influence each other towards H&S-related behaviour owing to the mobility of construction workers since co-workers build meaningful relationships among themselves and over a period may genuinely care about each other. According to Schwatka and Rosecrance (2016), when co-workers care about each other, they engage in H&S-related behaviour and may help each other in the event of an accident.

The authors postulate that the safety perception of a workgroup may be a major influence on the workers' safety participation. When co-workers interact in workgroups, they may exchange safety-related information (safety communication). This implies that management must ensure that workers are adequately empowered with the appropriate safety knowledge they need prior to the project commencement (Ajmal et al., 2020). Workers who do not know how to work safely will continue to increase the likelihood of hazards and accidents occurring in the workplace. The importance of safety training in workgroup safety commitment cannot be overemphasised. Management must consider this as a constant priority; workers who have a commensurate H&S knowledge are more confident about correcting their co-workers and teaching them how to work safely. This category of workers is also confident enough to discuss with their supervisors if they perceive that they know a safer way of doing their jobs which may eventually save management resources such as time and money. However, it is important that supervisors ensure that safer work methods as proposed by workers are indeed safe and will not jeopardise the attainment of the project's H&S-related goals. Workers committed to safety are also able to identify hazards in the workplace or work conditions of which management may be unaware, given that the workers are at the coalface. This implies that the management must establish an effective safety communication system in the workplace to encourage workers' safety voice. A work environment which has a complex reporting culture may indicate that workers are unable to voice safety concerns or that the voiced concerns may not reach the attention of management.

Workers' perception of the amount of the risk they are exposed to in their job roles may also impact on their compliance levels. When directly exposed to risk, workers are known to become more compliant to H&S regulations while they may not be overly concerned about compliance when they are convinced that the likelihood of accident occurrence is minimal (Xia et al., 2020). This again emphasises the need for improved safety training and education levels by management to ensure that, irrespective of the workers' perception of risk exposure, they continue to engage in H&S-related behaviour. In another study, Cohen (2009) highlighted the factors such as organisational safety justice and workers' values in safety commitment. Workers who perceive that they have been treated unfairly by management may show low levels of safety commitment. Similarly, workers' value systems may also influence their perception of safety commitment to the organisation, given that commitment as a concept is a display of an intention to engage in a long-lasting and beneficial relationship with an entity or an organisation. Therefore, while there is much emphasis on management's visible safety commitment and its influence on workers' perception of the same, it is important that stakeholders become aware of other equally important factors which may influence workers' H&S-related behaviour.

Workers at different levels require support from their fellow workers. The literature shows that workers who enjoy support from the organisation, supervisors and co-workers are more committed to workplace H&S compliance (Puah et al., 2016). The concept of co-worker support has the potential to improve H&S performance. Guchait et al. (2014) define co-worker support as the confidence in a fellow worker's willingness to assist in work-related responsibilities such as knowledge sharing and emotional support, among others. Owing to the uniqueness of the construction process, workers are expected to work in teams, depending on the size and complexity of the project. Subsequently, the interaction among workers often results in the creation of personal relationships that sometimes may extend beyond the workplace. Conchie et al. (2013) affirmed that a supportive work environment results in supportive workgroups. Moreover, co-worker support can originate from management to supervisors, supervisors to supervisors, supervisors to workers and workers to workers. However, workers who do not work together, despite being in the same organisation, may not share support owing to their lack of interaction, implying that workers' perception of support from their co-workers may influence H&S-related commitment.

6.5 Safety Commitment in Developing Countries

The reported findings present management's perception of workers' safety commitment in Nigeria and South Africa using the management safety commitment dimension of the NOSACQ-50. Management in this context comprised professionals such as architects, quantity surveyors, project managers, civil engineers, H&S managers and builders. Both countries were seen to perform well on the NOSACQ-50 interpretation scale. This indicates that the study group perceives that there is a commendable safety commitment at the group level. The measured dimension also relates to workgroups' shared behavioural norms and how these influence their H&S-related behaviour. There seems to be a good relational cohesion among workgroups which must be maintained. However, this may not be a strong indication of a serious safety commitment but rather a commendable social cohesion among workers, which may be explored for the benefit of improved safety commitment.

This study raises the issue that low levels of management safety commitment may not affect workers' social cohesion even if it does impact on their safety compliance levels. Workers may develop a 'we' versus 'them' attitude and look out for themselves irrespective of the management's H&S-related behaviour. The support that co-workers enjoy from their workgroups has been shown to boost their individual self-esteem, confidence and job satisfaction levels. Attiq et al. (2017) also mentioned that this may stimulate H&S-related behaviour. According to Choi et al. (2017), the complexities involved in construction projects often make it cumbersome to understand the workgroups' norms and social identities. For example, construction workers are mostly recruited temporarily on a project-to-project basis and their identities may only be reflective of the present work environment. Generally, Choi et al. (2017) indicated that the expectations of mid-level managers (supervisors) regarding the workgroup H&S standards may differ from what the workgroup perceives as acceptable standard, especially if the supervisor fails to communicate the same to the workgroup. Törner and Pousette (2009) posited that factors such as the project characteristics and the nature of the work; organisational structure; collective norms, values and behaviours and individual competencies and attitudes can influence the workgroup safety standards. Supervisors' expectation of H&S-related behaviour when communicated to workers often motivates them to work closely together, thereby improving their perception of management safety commitment and workgroup social cohesion. Organisational H&S standards are often set by management in line with statutory regulations to ensure that the H&S of workers, the safety of equipment, work conditions, process and activities are also as laid down (Zaira & Hadikusumo, 2017).

6.6 Conclusion

This chapter explored the concept of safety commitment from both management and workers' perspective and highlighted that though top management's visible safety commitment has been shown to reflect in workers' safety behaviour, management must note other factors which may influence workers' safety commitment. Safety performance in the workplace may be improved when management invests in improving workers' safety competence levels. An improvement in safety compliance and participation results in commendable safety performance in the workplace. Management safety commitment may not influence workgroup social cohesion, as workers may often find ways to look out for themselves, especially when they perceive that management has shown low interest levels in their collective and individual H&S. Safety competence must extend to supervisors as they are a direct and closer representation of management to the majority of workers. Management must ensure that supervisors represent the organisation in a manner which aligns with the organisation's stand on H&S compliance.

References

Ajmal M, Isha ASN, Nordin SM, Kanwal N, Al-Mekhlafi A-BA and Naji GMA (2020) A conceptual framework for the determinants of organizational agility: Does safety commitment matters? *Solid State Technology* 63(6). SST: 1–4.

Al-Bsheish M, bin Mustafa M, Ismail M, et al. (2019) Perceived management commitment and psychological empowerment: A study of intensive care unit nurses' safety. *Safety Science* 118(May). Elsevier: 632–640. http://doi.org/10.1016/j.ssci.2019.05.055

Amponsah-Tawiah K and Mensah J (2016) Occupational health and safety and organizational commitment: Evidence from the Ghanaian mining industry. *Safety and Health at Work* 7(3). Elsevier: 225–230. http://doi.org/10.1016/j.shaw.2016.01.002

Attiq S, Wahid S, Javaid N, et al. (2017) The impact of employees' core self-evaluation personality trait, management support, co-worker support on job satisfaction, and innovative work behaviour. *Pakistan Journal of Psychological Research* 32(1). National Institute of Psychology: 247–271.

Choi B, Ahn S and Lee S (2017) Construction workers' group norms and personal standards regarding safety behavior: Social identity theory perspective. *Journal of Management in Engineering* 33(4). ASCE: 1–11. http://doi.org/10.1061/(ASCE)me.1943-5479.0000511

Cohen A (2007) Commitment before and after: An evaluation and reconceptualization of organizational commitment. *Human Resource Management Review* 17(3). Sage Publications: 336–354. http://doi.org/10.1016/j.hrmr.2007.05.001

Cohen A (2009) A value based perspective on commitment in the workplace: An examination of Schwartz's basic human values theory among bank employees in Israel. *International Journal of Intercultural Relations* 33(4). Elsevier: 332–345. http://doi.org/10.1016/j.ijintrel.2009.04.001

Conchie SM, Moon S and Duncan M (2013) Supervisors' engagement in safety leadership: Factors that help and hinder. *Safety Science* 51(1). Elsevier: 109–117. http://doi.org/10.1016/j.ssci.2012.05.020

Dearmond S, Smith AE, Wilson CL, et al. (2011) Individual safety performance in the construction industry: Development and validation of two short scales. *Accident Analysis and Prevention* 43(3). Elsevier: 948–954. http://doi.org/10.1016/j.aap.2010.11.020

Emuze F and Smallwood J (2017) *Valuing people in construction*. London: Routledge. http://doi.org/10.4324/9781315459936

Geldart S, Smith CA, Shannon HS, et al. (2010) Organizational practices and workplace health and safety: A cross-sectional study in manufacturing companies. *Safety Science* 48(5). Elsevier: 562–569. http://doi.org/10.1016/j.ssci.2010.01.004

Griffin MA and Hu X (2013) How leaders differentially motivate safety compliance and safety participation: The role of monitoring, inspiring, and learning. *Safety Science* 60. Elsevier: 196–202. http://doi.org/10.1016/j.ssci.2013.07.019

Griffin MA and Neal A (2000) Perceptions of safety at work: A framework for linking safety climate to safety performance, knowledge, and motivation. *Journal of Occupational Health Psychology* 5(3). American Psychological Association: 347–358. http://doi.org/10.1037/1076-8998.5.3.347

Guchait P, Paşamehmetoğlu A and Dawson M (2014) Perceived supervisor and co-worker support for error management: Impact on perceived psychological safety and service recovery performance. *International Journal of Hospitality Management* 41. Elsevier: 28–37. http://doi.org/10.1016/j.ijhm.2014.04.009

Hardison D, Behm M, Hallowell MR, et al. (2014) Identifying construction supervisor competencies for effective site safety. *Safety Science* 65. Elsevier: 45–53. http://doi.org/10.1016/j.ssci.2013.12.013

He C, Jia G, McCabe B, et al. (2019) Impact of psychological capital on construction worker safety behavior: Communication competence as a mediator. *Journal of Safety Research* 71. Elsevier: 231–241. http://doi.org/10.1016/j.jsr.2019.09.007

Huang Y-H, Lee J, McFadden AC, et al. (2017) Individual employee's perceptions of 'group-level safety climate' (supervisor referenced) versus 'organization-level safety climate' (top management referenced): Associations with safety outcomes for lone workers. *Accident Analysis and Prevention* 98. Elsevier: 37–45. http://doi.org/10.1016/j.aap.2016.09.016

Ismail F, Ahmad N, Afida N, et al. (2012) Assessing the behavioural factors' of safety culture for the Malaysian construction companies. *Procedia – Social and Behavioral Sciences* 36(June 2011). Elsevier: 573–582. http://doi.org/10.1016/j.sbspro.2012.03.063

Kapp EA (2012) The influence of supervisor leadership practices and perceived group safety climate on employee safety performance. *Safety Science* 50(4). Elsevier: 1119–1124. http://doi.org/10.1016/j.ssci.2011.11.011

Kvalheim SA and Dahl Ø (2016) Safety compliance and safety climate: A repeated cross-sectional study in the oil and gas industry. *Journal of Safety Research* 59. Elsevier: 33–41. http://doi.org/10.1016/j.jsr.2016.10.006

Lawani K, Hare B, Cameron I and Sharon D (2018) Management's 'genuine benevolence' and worker commitment to health and safety – a qualitative study. In: *Proceedings of the joint CIB W099 and TG59 conference coping with the complexity of safety, health, and wellbeing in construction*, Salvador, Brazil, 1–3 August 2018, pp. 318–326. Available from: https://www.irbnet.de/daten/iconda/CIB_DC31539.pdf

Meyer JP and Allen NJ (1991) A three component conceptualization of organizational commitment. *Human Resource Management Review* 1(1). Elsevier: 61–89.

Pinion C, Brewer S, Douphrate D, et al. (2017) The impact of job control on employee perception of management commitment to safety. *Safety Science* 93. Elsevier: 70–75. http://doi.org/10.1016/j.ssci.2016.11.015

Puah LN, Ong LD and Chong WY (2016) The effects of perceived organizational support, perceived supervisor support and perceived co-worker support on safety and health compliance. *International Journal of Occupational Safety and Ergonomics* 22(3). Taylor and Francis: 333–339. http://doi.org/10.1080/10803548.2016.1159390

Reb J, Narayanan J and Chaturvedi S (2014) Leading mindfully: Two studies on the influence of supervisor trait mindfulness on employee well-being and performance. *Mindfulness* 2014(5). American Psychological Association: 36–45. http://doi.org/10.1007/s12671-012-0144-z

Schwatka NV and Rosecrance JC (2016) Safety climate and safety behaviors in the construction industry: The importance of co-workers commitment to safety. *Work* 54(2). PubMed: 401–413. http://doi.org/10.3233/WOR-162341

Törner M and Pousette A (2009) Safety in construction – a comprehensive description of the characteristics of high safety standards in construction work, from the combined perspective of supervisors and experienced workers. *Journal of Safety Research* 40(6). Elsevier: 399–409. http://doi.org/10.1016/j.jsr.2009.09.005

Wei J, Chen H and Qi H (2015) Who reports low safety commitment levels? An investigation based on Chinese coal miners. *Safety Science* 80. Elsevier: 178–188. http://doi.org/10.1016/j.ssci.2015.07.024

Xia N, Xie Q, Hu X, et al. (2020) A dual perspective on risk perception and its effect on safety behavior: A moderated mediation model of safety motivation, and supervisor's and coworkers' safety climate. *Accident Analysis and Prevention* 134(105350). Elsevier: 1–12. http://doi.org/10.1016/j.aap.2019.105350

Yule S, Flin R and Murdy A (2007) The role of management and safety climate in preventing risk-taking at work. *International Journal of Risk Assessment and Management* 7(2). Inderscience: 137–151. http://doi.org/10.1504/IJRAM.2007.011727

Zaira MM and Hadikusumo BHW (2017) Structural equation model of integrated safety intervention practices affecting the safety behaviour of workers in the construction industry. *Safety Science* 98. Elsevier: 124–135. http://doi.org/10.1016/j.ssci.2017.06.007

Zohar D (1980) Safety climate in industrial organizations: Theoretical and applied implications. *Journal of Applied Psychology* 65(1). American Psychological Association: 96–102. http://doi.org/10.1037/0021-9010.65.1.96

Zwetsloot GIJM, Kines P, Ruotsala R, et al. (2017) The importance of commitment, communication, culture and learning for the implementation of the Zero Accident Vision in 27 companies in Europe. *Safety Science* 96. Elsevier: 22–32. http://doi.org/10.1016/j.ssci.2017.03.001

7 Safety Communication

7.1 Introduction

Communication has been described as the most important factor in human relationships. In work organisations, it is described as the main driver of work-related outcomes. Without an efficient and effective communication network in an organisation, it becomes difficult to execute job-related tasks. Communication has been described in some quarters as the *social glue* in an organisation and the *nervous system* that holds workgroups; communication is vital that you *cannot not communicate* (Cacciattolo, 2015). It has been defined by many scholars and what seems to be a commonality in many definitions is that communication is an exchange of information and ideas among parties in a manner that is readily understandable. Rajhans (2009) mentioned that communication is the heart of an organisation, adding that human relationships or interactions are built on effective communication.

There are two main types of communication in the workplace, namely formal and informal types of communication. Formal communication mostly relates to management communicating to workers and workers to management (officially). It influences safety-related behaviour among workers. Informal communication takes place among workers. Both forms of communication may influence safety-related behaviour in an accident-prone workplace such as construction. This type of communication may reveal the strength of relationships which exist between workers and has a strong influence on the safety climate. In today's complex work environment, involving persons from different cultures, race and orientation, workers are constantly called upon to improve their communication skills in order to ensure that work-related outcomes are achieved efficiently and effectively. Communication should facilitate all the complexities that workers have to manage in the workplace, making the process seamless. This chapter explores in detail the importance of communication in the safety climate and how effective communication may aid the attainment of safety-related outcomes. It further elucidates the role of communication in H&S training and education to ensure that resources invested in improving workers' H&S knowledge also yields the anticipated results.

DOI: 10.1201/9781003361640-7

7.2 Safety Communication

The attainment of improved safety performance in an accident-prone workplace requires that all factors which contribute to the safety climate are carefully analysed for effective implementation. One of these factors is safety communication. Huang et al. (2018) described safety communication among supervisors as the extent to which the supervisor is able to pass safety-related information among workers in an understandable manner. Given that the supervisor is a level of management, this can be termed management safety communication. As important as safety communication is to climate studies, it will appear that there is still a dearth of literature focusing on the relationship between safety communication and safety performance in construction.

Effective safety communication must be two ways in the sense that the safety information being passed on must be understandable to the receiver. This implies that the responsibility of ensuring that safety communication takes place successfully rests with the sender, while it is the responsibility of the receiver to seek further clarification concerning any unclear information. This demonstrates that effective communication is an equal responsibility. In the construction workplace, the sender of safety information in most cases is either management or the supervisor and the receiver is the worker. A measurement of successful or efficient safety communication is observed in visible safety behaviour demonstrated by the worker. Unless this happens, it may be assumed that safety communication has not occurred. Huang et al. (2018) also mentioned that supervisors who continuously engage in safety communication with their workers have fewer incidents and accidents and higher levels of safety performance.

Safety communication requires skill. Safety leaders with poor communication skills will often experience levels of frustration as they are unable to present the 'bigger picture' of organisational safety to the workers. Until workers are able to understand the organisational safety policies and H&S regulations, safety leadership will continue to struggle in communicating this information to them (Rajhans, 2009). Lingard et al. (2019) confirmed that safety communication occurs between workers to managers (vertically) and among work peers (horizontally). Furthermore, they added that safety communication informs workers of the hazards and risks they are exposed to in their jobs and how to work safely. It also seeks specific information regarding the workers' experience and concerns and allows room for suggestions on how to improve H&S-related behaviour in a 'joint problem-solving' approach. A joint problem-solving approach is one which suggests that both workers and supervisors or management come together to assess safety-related challenges in the workplace and collectively seek methods which help solve the problems.

Safety communication involves the dissemination of safety-related information among workers in the most appropriate manner and is often a function of effective safety leadership. Kines et al. (2010) found that during the construction process, there are often verbal exchanges both horizontally and vertically. Communication between foremen and workers is mostly productivity-oriented but was improved in favour of H&S when feedback-based coaching regarding safety performance was introduced. This means that not all communication on construction worksites may be safety-inclined; nevertheless, adding safety dimensions to communication may result in safety outcomes (Guchait et al., 2014).

Certain conditions may impact on how workers communicate in the workplace. One notable is the concept of homophily. Studies in biological sciences have explained that it is the tendency of specific individuals to interact closely because they share peculiar characteristics such as gender, language, age, race and/or values. It also informs how individuals dissolve interpersonal relationships (Fu et al., 2012). Allison and Kaminsky (2016) add that unilingual construction work crews had above-average safety performance ratings when compared to multilingual crews. This emphasises the importance of selecting an appropriate language of instruction for the purpose of safety communication. According to Hargie (2016), findings from an earlier survey suggested that management perceived that they gave valuable feedback to their workers using appropriate communication methods. The workers, however, contended that this was not the case and that management needs to treat them with value, acknowledging the effort they put into their job roles in order to motivate them towards improved productivity. This may suggest that communication in such workplaces may not be effective and needs to be improved. Hargie further described the key elements to be considered in order to achieve effective communication as follows:

1   **Communicators**. These are the people involved in the communication process such as leaders and the workers with their individual attributes such as gender, height and attractiveness:

    a   Gender. Women may face different challenges in effective communication compared to men. This is related to the fact that most work environments globally are male-dominated. In construction there is an abundance of literature which supports this claim; for example, a female leader who needs to communicate to a workgroup of male workers may have to employ more skill and knowledge into effective communication.

    b   Height. A leader's height may influence the workers' perception of the power and authority they exert. This is owing to the fact that persons who lead children are often adults who are taller and children grow into adulthood with a mental picture of what a leader looks like. The link between height and power suggests that taller people are more positively regarded as leaders compared to shorter people.

    c   Attractiveness. Leaders who tend to have attractive physical attributes are often perceived to be more *confident, credible, persuasive, popular, trustworthy, likeable, outgoing, interesting, happy* and *intelligent*. This is a stereotype which has been widely discussed in scholarly works.

2   **Goals**. Organisational goals must drive the communication network to the point that people subconsciously become channelled and harnessed into pursuing the same goals.

3   **Messages**. In the dissemination of messages in an organisation, they must be credible, timely, clear, understandable without any form of ambiguity, accurate, consistent, not overwhelming, not too little information and relevant to the recipient. Workers try to make sense or meaning from messages they receive from management and the meaning may be lost in translation if not properly construed.

4   **Channel**. This refers to the medium adopted for communication which could be written, verbal, face to face, visual or audio. Findings have shown that workers prefer the face-to-face channel as it gives them some sense of value knowing that top management has chosen to communicate with them directly. Adopting channels such as email or memos to communicate with workers, especially in construction settings, may yield low results.

5   **Feedback**. This is a mechanism which allows for the sender of information to measure the extent to which the sent message was understood by the recipient. Feedback opens up the lines of communication from one-way to two-way communication and allows for shared understanding and mutuality in meaning.

6   **Context**. This refers to the sender's ability to recognise the circumstances which inform the need to share information and to ensure that the message sent aligns with the need.

Irrespective of how noble or well-crafted an organisational safety policy is, it will yield zero results if not effectively communicated to the workers. It may seem that low levels of safety communication may be related to the level of safety performance being observed by scholars. In many instances, management is aware of the importance of H&S-related behaviour among workers, and in some project sites, there is evidence that workers even memorise the safety regulations; however, this has still not translated into the expected results. Could it be that effective safety communication is lacking among supervisors and managers? Zwetsloot et al. (2017) in their investigation of factors such as safety communication in the implementation of a zero-accident vision revealed that safety communication levels in construction organisations were lower compared to other studied accident-prone organisations. In the construction industry, communication between management and management was slightly above average and that between management and worker somewhat lower. They further revealed that there is huge potential for management to invest in an effective supervisor safety communication strategy for the attainment of an improved H&S-related behaviour among workers. Discussion approaches to safety communication such as toolbox talks, safety walks and workshops have been found to contribute immensely to safety communication as they create a sense of openness and trust between supervisors and workers.

7.2.1 Open Safety Communication

Workers and management collectively finding solutions to safety-related challenges in a construction workplace is the outcome of an open communication system. This implies that it may not be achieved in a workplace which has not allowed open communication to grow. Open safety communication implies to an atmosphere where colleagues can be open with each other about their safety-related mistakes and concerns for the purpose of correction and learning (Zwetsloot et al., 2017). In the construction workplace, communication may take multiple forms. Jitwasinkul and Hadikusumo (2011) indicated in their report that the use of safety signs and symbols is often adopted in construction as a form of communication. However, workers have an indicated stronger preference for verbal communication. This may further imply that the workers are better motivated towards H&S compliance when they sense that they are acknowledged by their supervisors in a form of verbal communication. Hare et al. (2013), however, argued that the use of pictorial methods of communication in the construction workplace is more effective, given that the majority of workers have low levels of formal education. Moreover, in cases where workers are migrants or there is a language barrier between workers and the supervisor/management, verbal communication may not yield the desired results. Open safety communication not only influences workers' H&S-related behaviour, but it also holds management accountable for doing what the safety policy and regulations lay down (Zou et al., 2017). For example, open safety communication improves workers' safety voice; therefore, if workers are fully aware that they need a piece of equipment to work safely, they are able to insist that management improves their work conditions by holding them accountable.

7.2.2 Perceived Management Openness

Safety leadership has been proven to drive H&S-related behaviour in workplaces such as construction. Griffin and Hu (2013) mentioned that Hofmann and Morgeson's social exchange theory comes into play for improving H&S-related behaviour among workers. The theory states that employees who enjoy high relational quality with their managers are more likely to raise safety-related concerns and are more passionate about H&S in their workplace. Furthermore, Ng and Feldman (2012) added that human relationships evolve over time and the outcome of the evolution will often be determined by how parties to the relationship abide by their rules of exchange.

This introduces the rule of reciprocity; people who perceive that they are respected in a relationship are naturally inclined to accord the other similarly and vice versa. Subhakaran and Dyaram (2018), in line with the law of reciprocity, posited that workers who perceive that their managers value their contribution and genuinely care about their well-being are better placed to reciprocate in desired work behaviours. Nevertheless, owing to existing work hierarchies (among other factors), workers do not challenge the existing status quo for the fear of being perceived as self-promoting or critical. Detert and Burris (2007) described managerial openness as a 'subordinate's perception that their boss listens to them, is interested in their ideas, gives fair consideration to the ideas presented and at least, takes action to address any matters raised'. In work environments such as construction, management must ensure that relationships or social exchanges between them and workers enable feedback (two-way exchange) and a re-evaluation of the system.

Kaskutas et al. (2013) found in their study that foremen on construction projects sites received low levels of safety training and were unable to communicate safety requirements effectively to apprentice workers. As a result, this category of workers did not know what to do when faced with fall-related hazards. They reported that many manual construction workers join the industry with no formal construction-related training or prior experience. It is strongly recommended that an effective safety communication approach be adopted for the safety training of workers in construction. Zamani et al. (2020) mentioned that safety communication has the potential to provide an open communication and discussion platform where all workers, irrespective of cadre, may participate and improve their safety knowledge levels. Management must support H&S-related behaviour among individual workers and implement an effective safety training and education system, given that most general (manual) workers have low levels of formal safety education.

7.3 Safety Training and Education

Safety communication is the exchange and sharing of H&S-related knowledge among members in an organisation or workgroup to improve their H&S-related behaviour, safety compliance and participation as well as their job risk-perception levels. The acquisition of safety knowledge is achieved through safety training, according to Liu et al. (2018), which significantly improves construction workers' hazard recognition and risk perception. Zamani et al. (2020) indicated that safety training may adopt a formal or informal approach. Safety training through formal communication methods refers to the use of specific channels of interaction with management or supervisors such as safety training, issuing of work orders, written notifications, memos, emails, safety signs and toolbox talks. Management also issues safety work orders such as specific policies, procedures and instructions so that workers can follow with the perception that the approach improves safety compliance and related behaviour among workers. Toolbox talks or meetings are a platform where workers meet to discuss safety-related matters prior to the commencement of daily tasks. Efficient safety training has been shown to be an essential element in the implementation of successful safety management systems.

Informal communications in construction refers to the formation of a communication channel which does not conform to a particular systematic basis (Zamani et al., 2020). It usually takes the form of mentorship, informal discussions, peer-to-peer safety dialogue or the use of social media and instant messaging applications. These methods of communication have been found to yield positive results in the transfer of knowledge. For example, mentoring as a form of knowledge transfer is one of the oldest methods still being used to pass down professional knowledge, mostly from an older generation to a younger one. In workplaces such as construction, mentors are mostly older and train younger workers in accordance with how they were trained when they started out. There is empirical evidence supporting improved safety compliance and participation levels among older workers compared to younger workers. This implies that they naturally pass on safer work methods to new workers.

Reese (2016) confirmed that safety training and education are a sure approach towards accident prevention and insisted that it extends beyond only general workers to include forepersons, supervisors and managers. Reese highlighted that when conducting safety training programmes for new workers, the following subjects should be included:

- Company H&S programme and policy
- Employee and supervisors' responsibility
- Hazard communication training
- Location of first aid stations, fire extinguishers and emergency telephone numbers
- Site-specific hazards
- Procedures for reporting injuries
- Use of PPE
- Hazard identification and reporting procedures
- Review of each safety and health applicable to the job task
- Site tour or map where appropriate

Today's work environment is complex and continually evolving; periodically, newer and safer methods of carrying out hazardous tasks are introduced. On some occasions, innovation introduces more efficient safety gear, which workers will need to learn how to use. It is therefore important that no workers perceive that they are at a level where they no longer require safety training because it is by no means exhaustive. Construction workers must have a positive attitude towards continuous safety training; an open mind helps them to learn, unlearn and relearn. Continuous safety training brings construction workers up to speed and helps in hazard detection and the reduction of near misses. Management has often been accused of being more interested in production-related goals than the H&S of both workers and the workplace. However, although this does not always align with management's perception of their attitude towards workers' and workplace H&S (Skeepers & Mbohwa, 2015), an investment in safety training and worker education may yield improved safety performance results.

According to Alsharef et al. (2020), the following barriers exist regarding construction safety training:

- Transient nature of the construction workforce and temporal nature of projects discourage management from adopting more effective and resource-intensive training methods.
- Competing interest and the scarcity of resources.
- Difficulty in quantifying and communicating the tangible benefits of training.
- Lack of expertise among industry personnel in offering effective training.
- The perception that training increases the cost of doing business.

- Trained workers being poached by competing businesses.
- Time constraint and intensive project schedules.
- Lack of support from top management and supervisors.
- Disparity between the number of workers and qualified trainers in the industry.
- Mishandling of organisation's internal and external safety knowledge management resources.
- The belief among workers that they already possess the necessary skills partly due to the repetitive training on the same subject matter.
- Employers often unaware of the latest peer-reviewed training interventions published in journals.

Often it has been seen that management adopts methods which they find convenient or which they perceive are a good fit for the safety training of general construction workers. Research has shown that, owing to their low levels of formal education, workers find these approaches boring and may struggle to understand what is being taught. For example, the use of seminars, lectures and workshops for safety training among manual construction workers may not yield the anticipated results.

7.3.1 Construction Safety Training Methods

The following methods are recommended for construction safety training:

- **E-learning tools**. The use of Internet-based resources or local media storage for the purpose of safety training.
- **Peer-to-peer training**. Safety training between colleagues mostly in the same workgroups.
- **Mentoring**. Safety training conducted by an older colleague to a younger or new worker.
- **Game technology-based training**. The use of three-dimensional digital simulations to create a construction project site and simulate common or possible hazards and how the worker can work safely.
- **Safety campaigns**. The use of interactive fun methods for the communication of safety-related concerns to construction workers.
- **Visual aids**. The use of safety signs, symbols and pictures for safety training.
- **Modified instructor-led training**. The use of a classroom method for peer-to-peer instruction under the supervision of a higher cadre worker. This approach provides feedback to management on the training they have conducted with workers.

7.3.2 Safety Knowledge and Communication

The extent to which workers in the construction workplace understand the intricacies of H&S practices and procedures is referred to as their safety knowledge

(Shen et al., 2017). This is also found to predict safety compliance and participation which are both mediators in a safety climate (Jiang et al., 2010). Safety knowledge is often discussed along with safety motivation, which is described as the workers' inclination towards performing their jobs safely. Safety learning results in an improved safety knowledge which, according to Vinodkumar and Bhasi (2010), influences workers' perception of management's support towards their safety-related activities. Safety knowledge does not necessarily predict safety motivation; nevertheless, Loosemore and Malouf (2019) mentioned that factors such as the formal educational background of the workers may influence their grasp of safety learning programmes which focus on instructor-centric pedagogy. This approach may yield few positive results in older workers. Xu et al. (2019) indicated that factors such as work type, complexity, experience, age, motivation, emotion and learning strategy may impede the workers' learning abilities; however, the application of acquired knowledge may result in an improved adherence to H&S regulations and safety compliance. The adoption of learner-centric andragogy for the improvement and engagement of workers in H&S-related issues is highly encouraged.

The advent of technology in knowledge sharing has introduced more effective approaches which scholars such as Le et al. (2014) proposed have the potential to change the dynamics of safety knowledge sharing among construction workers. In their report, they introduced the use of a *social network system for sharing safety and health knowledge*. They contended that the proposed network allows for the uploading of H&S-related data and availing it on the public domain. This approach to H&S knowledge sharing has the potential of breaking down complex terminologies and concepts for construction workers and availing them of an opportunity to improve their H&S knowledge levels. In addition, proposals such as these may be translated into different languages, thereby removing the language-related challenges in safety training. Another advantage is that new workers interested in joining the construction industry may have an opportunity to study at their own pace before applying, thereby reducing accident levels often associated with apprentice construction workers. Construction managers are able to look up improved and safer methods of approaching a project of which they have little or no knowledge and are able to modify approaches to suit their own characteristics.

7.4 Safety Communication in Developing Countries

The data collected from the study conducted in Nigeria and South Africa show that there is a dominance of verbal safety communication methods in the form of instructions. This implies that communication in such workplaces may be one way (management to worker) and does not allow for a feedback mechanism. This also suggests that workers' safety voice levels may be low or non-existent, while open safety communication levels may also be low. Supervisors indicate that they also adopt the use of safety signposts placed at locations where workers can see these as they work. They also use instant messaging applications such as text messages and WhatsApp for the purpose of passing information. There is also evidence of toolbox meetings on sites. When asked about methods employed for the purpose of safety training, supervisors indicated that they used seminars, workshops and formal training. This implies that the knowledge level of appropriate communication channels and methods for the purpose of safety training is low even between management and supervisors. The use of an appropriate communication method, especially between management and frontline construction workers, may make the difference to the level of understanding achieved between parties and may also promote safety-related behaviour. This applies to the process of safety knowledge transfer between workers and management.

The aforementioned results underscore the need to mitigate poor safety communication, which is critical to promoting a safety climate. Pandit et al. (2019) maintained that a positive relationship exists between safety climate and safety communication levels. In a study that examined the factors that promote or impede effective safety communication, Pandit et al. (2019) found that construction crews that demonstrated higher levels of cohesion exhibited superior safety communication levels. In effect, evidence suggests that a synergy exists between safety climate and crew cohesion in improving safety communication levels (Pandit et al., 2019).

7.5 Conclusion

This chapter explored safety communication in light of safety training and education among management and general construction workers. The chapter highlighted the importance of safety communication in construction and how it can be improved to better serve the workers in construction. Safety communication is fundamental to outcomes to realise task completion without harming people in the frontline of operations. Poor safety communication, which remains a problem, should be addressed in developing countries by removing systemic barriers. One way to address the issues is using appropriate training methods. Safety training methods must be chosen on the basis of appropriateness and not based on the management or supervisor's convenience; the characteristics of general construction workers must inform the decision to choose which method. Safety training must also be continuous and sustained across all categories of workers.

References

Allison L and Kaminsky J (2016) Safety communication networks: Females in small work crews. *Journal of Construction Engineering and Management* 43(8). ASCE: 1–46. http://doi.org/10.1061/(ASCE)CO.1943-7862.0001344

Alsharef A, Albert A and Bhandari S (2020) Construction safety training: Barriers, challenges, and opportunities. In: *Construction research congress 2020: Safety, workforce, and education*, Reston, VA: American Society of Civil Engineers (ASCE), pp. 547–555. http://doi.org/10.1061/9780784482872.059

Cacciattolo K (2015) Defining organisational communication. *European Scientific Journal* 11(20). European Scientific Institute, ESI: 79–87.

Detert JR and Burris ER (2007) Leadership behavior and employee voice: Is the door really open? *Academy of Management Journal* 50(4). Academy of Management: 869–884. http://doi.org/10.5465/AMJ.2007.26279183

Fu F, Nowak MA, Christakis NA, et al. (2012) The evolution of homophily. *Scientific Reports* 2. Nature: 2–7. http://doi.org/10.1038/srep00845

Griffin MA and Hu X (2013) How leaders differentially motivate safety compliance and safety participation: The role of monitoring, inspiring, and learning. *Safety Science* 60. Elsevier: 196–202. http://doi.org/10.1016/j.ssci.2013.07.019

Guchait P, Paşamehmetoğlu A and Dawson M (2014) Perceived supervisor and co-worker support for error management: Impact on perceived psychological safety and service recovery performance. *International Journal of Hospitality Management* 41. Elsevier: 28–37. http://doi.org/10.1016/j.ijhm.2014.04.009

Hare B, Cameron I, Real KJ and Maloney WF (2013) Exploratory case study of pictorial aids for communicating health and safety for migrant construction workers. *Journal of Construction Engineering and Management* 139(7). ASCE: 818–825. http://doi.org/10.1061/(asce)co.1943-7862.0000658

Hargie O (2016) The importance of communication for organisational effectiveness. In: *Psicologia do trabalho e Das Organizações*. Braga, Portugal: Arizona, pp. 15–32. http://doi.org/10.17990/Axi/2016

Huang Y, Sinclair RR, Lee J, et al. (2018) Does talking the talk matter? Effects of supervisor safety communication and safety climate on long-haul truckers' safety performance. *Accident Analysis and Prevention* 117(September 2017). Elsevier: 357–367. http://doi.org/10.1016/j.aap.2017.09.006

Jiang L, Yu G, Li Y, et al. (2010) Perceived colleagues' safety knowledge/behavior and safety performance: Safety climate as a moderator in a multilevel study. *Accident Analysis and Prevention* 42(5). Elsevier: 1468–1476. http://doi.org/10.1016/j.aap.2009.08.017

Jitwasinkul B and Hadikusumo BHW (2011) Identification of important organisational factors influencing safety work behaviours in construction projects. *Journal of Civil Engineering and Management* 17(4). Taylor and Francis: 520–528. http://doi.org/10.3846/13923730.2011.604538

Kaskutas V, Dale AM, Lipscomb H, et al. (2013) Fall prevention and safety communication training for foremen: Report of a pilot project designed to improve residential construction safety. *Journal of Safety Research* 44(1). Elsevier: 111–118. http://doi.org/10.1016/j.jsr.2012.08.020

Kines P, Andersen LPS, Spangenberg S, et al. (2010) Improving construction site safety through leader-based verbal safety communication. *Journal of Safety Research* 41(5). Elsevier: 399–406. http://doi.org/10.1016/j.jsr.2010.06.005

Le QT, Lee DY and Park CS (2014) A social network system for sharing construction safety and health knowledge. *Automation in Construction* 46(October 2014). Elsevier: 30–37. http://doi.org/10.1016/j.autcon.2014.01.001

Lingard H, Pirzadeh P and Oswald D (2019) Talking safety: Health and safety communication and safety climate in subcontracted construction workgroups. *Journal of Construction Engineering and Management* 145(5). ASCE: 1–11.

Liu D, Gambatese JA and Jin Z (2018) Construction safety training methods and effectiveness for non-native workers. In: *Proceedings of the joint CIB W099 and TG59 conference coping with the complexity of safety, health, and wellbeing in construction*, Salvador, Brazil, 1–3 August 2018, pp. 140–148. Available from: https://www.irbnet.de/daten/iconda/CIB_DC31519.pdf

Loosemore M and Malouf N (2019) Safety training and positive safety attitude formation in the Australian construction industry. *Safety Science* 113(August 2018). Elsevier: 233–243. http://doi.org/10.1016/j.ssci.2018.11.029

Ng TWH and Feldman DC (2012) Employee voice behaviour: A meta-analytic test of the conservation of resources framework. *Journal of Organizational Behavior* 33. Wiley-Blackwell: 216–234. http://doi.org/10.1002/job

Pandit B, Albert A, Patil Y and Al-Bayati A (2019) Fostering safety communication among construction workers: Role of safety climate and crew-level cohesion. *International Journal of Environmental Research and Public Health* 16(1). MDPI: 1–16. http://doi.org/10.3390/ijerph16010071

Rajhans K (2009) Effective organizational communication: A key to employee motivation and performance. *Interscience Management Review* 2(2). Interscience: 145–149. http://doi.org/10.47893/IMR.2009.1040

Reese CD (2016) Who knows what: Safety and health training. In: *Occupational health and safety management: A practical approach*. 3rd ed. London: Taylor and Francis, pp. 245–258.

Shen Y, Ju C, Koh TY, et al. (2017) The impact of transformational leadership on safety climate and individual safety behavior on construction sites. *International Journal of Environmental Research and Public Health* 14(1). MDPI AG: 1–17. http://doi.org/10.3390/ijerph14010045

Skeepers NC and Mbohwa C (2015) A study on the leadership behaviour, safety leadership and safety performance in the construction industry in South Africa. *Procedia Manufacturing* 4(Iess). Elsevier BV: 10–16. http://doi.org/10.1016/j.promfg.2015.11.008

Subhakaran SE and Dyaram L (2018) Interpersonal antecedents to employee upward voice: Mediating role of psychological safety. *International Journal of Productivity and Performance Management* 67(9). Emerald Publishing: 1510–1525. http://doi.org/10.1108/IJPPM-10-2017-0276

Vinodkumar MN and Bhasi M (2010) Safety management practices and safety behaviour: Assessing the mediating role of safety knowledge and motivation. *Accident Analysis and Prevention* 42(6). Elsevier: 2082–2093. http://doi.org/10.1016/j.aap.2010.06.021

Xu S, Zhang M and Hou L (2019) Formulating a learner model for evaluating construction workers' learning ability during safety training. *Safety Science* 116(July 2019). Elsevier: 97–107. http://doi.org/10.1016/j.ssci.2019.03.002

Zamani V, Banihashemi SY and Abbasi A (2020) How can communication networks among excavator crew members in construction projects affect the relationship between safety climate and safety outcomes? *Safety Science* 128(104737). Elsevier: 1–12. http://doi.org/10.1016/j.ssci.2020.104737

Zou PXW, Lun P, Cipolla D, et al. (2017) Cloud-based safety information and communication system in infrastructure construction. *Safety Science* 98. Elsevier: 50–69. http://doi.org/10.1016/j.ssci.2017.05.006

Zwetsloot GIJM, Kines P, Ruotsala R, et al. (2017) The importance of commitment, communication, culture and learning for the implementation of the Zero Accident Vision in 27 companies in Europe. *Safety Science* 96. Elsevier: 22–32. http://doi.org/10.1016/j.ssci.2017.03.001

8   Safety Trust

8.1 Introduction

The concept of trust has always been approached from a social capital perspective established between individuals in a group or individuals to organisations.

> Trust has been defined as the willingness of a party to be vulnerable to the actions of another based on the expectations that one will perform what they have sworn or committed to perform to the trustor irrespective of the ability to monitor or control the other party.
> (Kożuch & Sienkiewicz-Małyjurek, 2022; Mehta et al., 2020)

Trust is vital to the sustenance of relationships at both interpersonal and organisational levels; it can be said in a business arrangement that trust is first what first endears customers to an organisation and then retains their loyalty. A breakdown in trust from an organisation may result in untold losses; this is what every business works assiduously to avoid. Trust facilitates communication and is important in conflict resolution. In safety-related studies, workers' trust in management has been found to be a contributor to their H&S-related behaviour (Gümüştaş & Küskü, 2021). When management proposes or commits to a safety-related outcome and fails to do as agreed, this behaviour creates a level of distrust among workers for management. Distrust is equally as important as trust in an organisational relationship. Roughly described, distrust is a negative breakdown of trust between entities in an agreement. Burt et al. (2009), however, argued that in a workgroup setting, some level of distrust may contribute to workgroup safety, implying that workers with low levels of trust among their colleagues may engage in personal H&S-related behaviour. This chapter considers safety trust from a construction perspective. Although there is insufficient literature on the topic in construction-related studies, this report nevertheless seeks to emphasise its importance in safety management.

8.2 Safety Trust in Construction

Safety trust may be described as the preparedness of both management and workers in an accident-prone organisation to be vulnerable to each other's actions on

DOI: 10.1201/9781003361640-8

the premise that management will deliver on workplace safety requirements and workers will comply with safety regulations. Gümüştaş and Küskü (2021) defined employee safety trust as safety-related attitudes and behaviours of members of an organisation which in turn affect safety performance. Safety trust can be viewed from a management-to-worker, worker-to-management and worker-to-worker perspectives. When management invests in worker-related safety outcomes, management trusts that workers behave in a manner which reflects their commitment; conversely, when management fails to uphold their safety-related commitment, then distrust between the two parties is created. The impact of distrust is captured by the attribution theory, which states that individuals, after learning of or experiencing an event which results in negative outcomes, may make attributions about its cause which further inform their emotional response and future behaviour towards it (Mehta et al., 2020).

Safety trust among co-workers is another interesting perspective. Researchers have demonstrated its relationship to safety compliance and participation in workplaces such as construction (Felletti & Paglieri, 2019). Tan and Lim (2009) described co-workers as individuals who hold equal power or are on the same level of authority and interact in the course of executing their duties in the workplace. Furthermore, they defined trust in co-workers as the willingness of such persons to become vulnerable to the actions of each other, even when they have no control over their behaviour. Erdem and Özen-Aytemur (2014) added that trust in co-workers can be context-specific, as in the case of organisational trust. They mentioned that in a work environment, trust is perceived as positive expectations that co-workers have concerning their collective intentions and behaviour, which are generally rooted in their organisational roles, relationships, experiences and interdependencies. According to Schwatka and Rosecrance (2016), owing to the uniqueness and peculiarities of work operations in the construction industry, such as varying job sites and subcontracting, co-workers may be disconnected from top management and more connected with their immediate workgroups. The workgroup's commitment to H&S may influence the co-workers' H&S-related behaviour. This implies that co-workers' trust may also be lateral (worker to worker) and horizontal (worker to management). Workers may define their safety behaviour around the level of trust they have in their co-workers' competence.

Ahmed et al. (2022) suggested that workers' safety trust may be related to the type of safety leadership adopted in the workplace. Specifically, they indicated that the servant leadership approach, which is service-oriented, has been associated with employee job-related outcomes such as engagement, performance, creativity and job satisfaction. This may not be too far from the social exchange theory, which indicates that there are benefits to be enjoyed in a social exchange between members of the same community. A servant leadership approach suggests that because employees perceive that their leader tries to serve them with regard to their H&S, they will put extra effort into their H&S-related behaviour. This category of workers, because they have high levels of trust in their leader, will be more creative towards safety participation and compliance. Furthermore, they may admit to having higher job satisfaction levels.

8.2.1 Antecedents of Safety Trust

- **Benevolence**. This is a display of goodwill and noble intentions towards the welfare of another. It is the desire of the leader to engage in positive behaviour in favour of the workers. Safety leaders high on benevolence attribute have been found to motivate and influence H&S-related behaviour seamlessly among workers because the workers trust their intentions.
- **Ability**. This is described as demonstrable skill and competence in a specific field of endeavour. H&S ability which emanates from the leader communicates to the workers that the leader is knowledgeable and is able to mentor them on safer work approaches.
- **Integrity**. This is seen in the manager's ability to uphold standards relevant to the workers. In the context of safety, it may be viewed as the ability of the safety leader to uphold H&S regulations and standards, which in turn presents the leader as being committed to the workers' H&S.

8.2.2 Factors That Influence Workers' Safety Trust in Management

According to Conchie et al. (2011), trust develops from an individual's belief of another's trustworthiness, given that in a construction safety context, workers' safety trust relates to *honesty, openness and concern for other's welfare and safety*. A determinant of safety trust in the workplace may also be a function of management's visible safety commitment to ensuring that safety performance in the workplace is not compromised. This approach by management may be displayed in the provision of PPE for workers, ensuring the work conditions to which workers are subjected comply with H&S regulations, and investing in workers' H&S knowledge. Workers who perceive that management goes above and beyond the statutory requirements to demonstrate their regard for employee H&S are easily won over regarding safety trust. The literature has revealed certain factors which influence workers' safety trust in management. These are discussed as follows:

- **Perceived status of the H&S officer**

According to Zohar (1980), while a strong management safety commitment ensures successful implementation of safety regulations, the rank and status of the H&S officers may further influence their ability to implement H&S management systems. Organisations with good H&S systems often have H&S officers who experience a higher status perception of workers (Mearns et al., 2003). The H&S officers are required to have the necessary skills, knowledge and experience for the role, given that they are mostly in direct contact with the workers and can influence their H&S-related behaviour (Zin & Ismail, 2012). Having an H&S officer who falls short of the minimum requirement is a poor representation of management in the eyes of both the H&S inspection officers and the workers. It is important that the H&S officers are able to carry themselves in a manner that suggests to the workers in particular that they are fully knowledgeable in their task and are ready to mentor workers who look up to them. The status of the H&S officer places them on a pedestal of authority and also improves their confidence as they interact with workers and influence H&S-related behaviour among workers (Vinodkumar & Bhasi, 2010). It is important to note here that H&S officers must demonstrate effective safety leadership and communication by treating workers with dignity and respect so that they are able to win their trust. A knowledgeable H&S officer is also able to provide relevant feedback to management regarding workers' H&S behaviour and to recommend an appropriate approach which addresses workers' safety motivation. The status of the H&S officer in turn communicates to the workers management's safety commitment and priority of their H&S (Zohar, 2000).

• **Frequency of H&S inspection**

H&S inspection is a management practice which aims to prevent or reduce to the barest minimum the occurrence of H&S-related incidents, accidents and near misses. It also enables the identification and assessment of hazards and provides a database from reported incidents for analysis of organisational learning and enables management to make data-informed decisions for the purpose of improved safety performance in the organisation (Hajmohammad & Vachon, 2014). Regularly carrying out H&S inspections in the workplace is an efficient H&S management practice as it exposes specific areas that need attention (Vinodkumar & Bhasi, 2010). H&S inspections also enable management to identify which risk is associated with which accident and the attending impact attached to a specific accident occurrence. For example, regular H&S inspections may help in terms of identifying the causes of falls and struck-by; analysis of these accidents can easily reveal a serious issue to be addressed (Chi & Han, 2013). Management must ensure that H&S inspections are carried out regularly as stipulated to further improve workers' perception of management safety priority and safety trust.

• **Workers' perception of H&S systems**

A systematic review was conducted by Aburumman et al. (2019) regarding H&S interventions targeted at improving safety performance in workplaces such as manufacturing, construction, utilities, forestry, healthcare, transportation, among others. It revealed that interventions are mostly targeted at workgroups, then organisational levels and then the individual. The most efficient interventions were those which focused on management's H&S priority, leadership style and safety behaviour. Zhou et al. (2015) indicated that workplace H&S studies focus on the characteristics of an individual worker such as behaviour, perception, attitude, competence and psychology seeing that they are direct stakeholders in workplace accidents. However, it is the workers' perceptions of group- or organisational-level H&S characteristics that are more fundamental, given that they can impact the workers' H&S-related behaviour in the workplace. Workgroups with positive shared perceptions of the safety system are more likely to engage in behaviour that promotes safety participation, which invariably results in improved H&S outcomes.

8.3 Workers' Trust in Safety Management Systems

Safety management system are mechanisms put in place in an accident-prone workplace, which ensure that the likelihood of hazards that lead to accidents are prevented or minimised as far as possible. According to Yiu et al. (2019), a safety management system is comprehensive and includes *policies, objectives, plans, procedures, organisations, responsibilities and other safety improvement measures*. Wu et al. (2019) highlighted that a system ideally consists of multiple intertwined components which work in synergy towards an anticipated goal or outcome. In complex safety systems such as construction, safety tenets must be identified through systematic thinking. The set objective in a safety system is the elimination of workplace hazards which increase the likelihood of accident occurrence. Yiu et al. (2018) mentioned that the implementation of a safety management system is necessary in accident prevention and reduction.

The literature reveals that there are several perspectives which must be considered in the implementation of safety management; these include human behaviour and safety management practices and how they both interrelate (Wu et al., 2019). The concept of safety trust flows from human behaviour and relationships. It is safe to assume here that workers who perceive that safety management systems established in the workplace are implemented optimally may have high safety trust levels. Skeepers and Mbohwa (2015) contended that two-way open communication in the workplace between management and workers builds trust and promotes a positive safety climate and culture. Regarding systems thinking in establishing an effective safety management system, Brovedani (2020) emphasised the importance of budgeting for safety-related outcomes right from the planning stages of the project. A healthy budget devoid of bureaucracy makes it easier for workers to be equipped with the skills, knowledge and equipment they need to manage safety-related challenges at work.

8.4 Safety Distrust in Construction

The theory of attribution aids in the understanding of distrust in an organisational setting.

Mehta et al. (2020) indicated that when an event occurs which leads to lower trust levels, the trustor assesses the events on three attribution levels:

- First is the locus of casualty (internal or external), which refers to the perception of the cause of an event being either within the control of the perpetrator or beyond their control.
- Second is controllability, which refers to who is taking responsibility for the outcome: Is it the organisation or they are transferring accountability or blame to another entity?

- Third is stability, which has to do with how often related or similar outcomes occur in the organisation.

Workers' distrust in a safety system has been shown to contribute to negative adverse safety outcomes such as CWB. The literature on safety distrust suffers a similar fate as safety trust in terms of low interest levels among scholars, resulting in the absence of sufficient literature. Gümüştaş and Küskü (2021) stated that distrust breeds where one party perceives that there may be incongruence in value or that others have failed to accept the trustee's core values they perceive as incompatible with theirs. However, they further highlighted the following as antecedents of safety distrust:

- **Malevolence.** This refers to hostile behaviour perpetuated with the sole aim of causing harm to another. A leader who exhibits high levels of this behaviour must never be allowed in an accident-prone workplace. High levels of distrust may lead to malevolent behaviour among workers (Conchie et al., 2011). Management must ensure that safety trust in the workplace is maintained.
- **Incompetence.** When leaders demonstrate low levels of safety knowledge, workers perceive them as incompetent. This perception among workers may be difficult to change when established. Management must ensure that their representatives display high levels of safety awareness in order to win the trust of workers.
- **Deceit.** This is a poor representation of management which in this case may relate to having a supervisor or leader whose behaviour is contrary to what the organisation represents in terms of H&S.

Safety distrust is often viewed as an opposite of safety trust and is easily associated with negative outcomes. However, Conchie et al. (2011) argued that safety distrust has been seen to yield some positive outcomes in that it has the capacity to increase vigilance and wariness. These are necessary ingredients in building a healthy and resilient organisation. It appears that safety distrust has more adverse effects on human interaction. When it abounds, other safety-related outcomes in the workplace, such as safety communication and incident reporting, may be impacted. In an earlier study, Burns et al. (2006) concluded that trust and distrust are two distinct constructs which can exist simultaneously among workers. When management consistently takes immediate action regarding incident reports and engages in effective open safety communication with the workers on how to improve workplace H&S, safety trust is eventually established (Conchie & Burns, 2008).

8.5 Benefits of Safety Trust

Safety trust among members in the same workplace has been shown to yield positive organisational outcomes. The safety trust mostly resulted from the members' social interaction and understanding of each other's expectations and approach to work. Mehta et al. (2020) identified the following outcomes of safety trust (Table 8.1):

Table 8.1 Safety trust benefits.

Establishment of integrative work practices	Workers who have worked together closely and established levels of trust among themselves are able to set up work practices which prevent certain difficulties that may arise. This is because they know their various strengths and abilities and are able to harness these for the greater good of their work goals and enjoy a feeling of collective responsibility.
Common philosophy	This is the ability of the workgroup to unify their thoughts and behaviour regarding work. They are able to build values, mutual respect and understanding on how they expect to be treated and what they will not accept. They are also able to agree on work norms and behaviour.
Open communication	Safety trust builds a work environment which upholds safety voice where workers are able to express their safety-related concerns without the fear of prejudice, intimidation or harassment. Workers are able to speak up when they perceived that their work environment puts their H&S at risk and may refuse to work until their concerns have been addressed by management, thereby improving accountability levels in the workplace. This type of workplace facilitates safety knowledge transfer among workers through experience sharing and also encourages incident reporting for organisational learning.
Clear role expectations	Given that workers have a fair understating of each other's capabilities, they are able to manage their work-related expectations effectively. Roles in the workgroup are assigned according to each worker's strengths.
Shared trust climate	This implies that workgroup members trust and are loyal to the team. Therefore, if H&S behaviour is enshrined in such teams, they are all confident of a commendable safety compliance and participation level even without being monitored or controlled.

8.6 Safety Trust in Developing Countries

The reported study investigated workers' perception of reliability of the H&S system and how confident workers are regarding the safety management in the workplace from two cohorts in Nigeria and South Africa. Findings suggest that participants from both cohorts have high levels of perception relating to reliability and confidence in the safety management practices in their workplace. Participants from both cohorts also demonstrated that their trust in the efficacy of the safety system was very good when assessed on the NOSACQ-50 interpretation scale. This result may be related to the fact that workers, irrespective of their cadre, still perceive themselves as a workgroup. This confirms the social exchange theory which emphasises that close relations between workers over time builds trust among themselves. It is important to note that this trust may be explored in favour of improved safety performance for the organisation as it becomes easier to motivate workers towards safety compliance and participation.

The results from Nigeria and South Africa resonate with previous studies regarding the role of trust in fostering the correct safety climate within a group. Kath et al. (2010) examined a hypothesis which investigated whether job safety relevance would moderate the relationship between safety climate and trust using survey research conducted with 599 employees from 97 workgroups. Although not a construction-related cohort, the perceptions of the respondents in the service industry indicated that trust mediates the relationship between safety climate and organisational outcomes. In addition, Kath et al. (2010) reported that the relationship between safety climate and trust was stronger within workgroups where safety was more relevant.

8.7 Conclusion

This chapter argues that safety trust leads to improved safety performance. The subject of safety trust was also assessed alongside safety distrust. Though scholars may be divided with regard to safety trust and distrust, either being two ends of the same concept or two varying concepts, it is clear that in the safety literature, an increased level of safety trust results in a decrease in safety distrust. Management is encouraged to prioritise safety, ensuring that there is effective safety leadership which upholds H&S compliance and participation to an unfaultable degree as this communicates a high level of safety commitment and priority among workers. This results in improved safety trust among management, workers and workgroups. A safety climate where trust exists will have the right management attitudes towards H&S, bidirectional safety communication and motivation, to mention but a few.

References

Aburumman M, Newnam S and Fildes B (2019) Evaluating the effectiveness of workplace interventions in improving safety culture: A systematic review. *Safety Science* 115(January). Elsevier: 376–392. http://doi.org/10.1016/j.ssci.2019.02.027

Ahmed F, Xiong Z, Faraz NA and Arslan A (2022) The interplay between servant leadership, psychological safety, trust in leader and burnout: Assessing causal relationships through a three-wave longitudinal study. *International Journal of Occupational Safety and Ergonomics*. Taylor and Francis: 1–44. http://doi.org/10.1080/10803548.2022.2086 755. Available from: https://www.tandfonline.com/doi/abs/10.1080/10803548.2022.20 86755?journalCode=tose20

Brovedani L (2020) A matter of trust: How to evolve and manage worker health and safety. *Environmental Health and Safety Today (EHS Today)*. Available from: https://www.ehstoday.com/safety-leadership/article/21120116/a-matter-of-trust-how-to-evolve-and-manage-worker-health-safety

Burns C, Mearns K and McGeorge P (2006) Explicit and implicit trust within safety culture. *Risk Analysis* 26(5). PubMed: 1139–1150. http://doi.org/10.1111/j.1539-6924.2006.00821.x

Burt CDB, Chmiel N and Hayes P (2009) Implications of turnover and trust for safety attitudes and behaviour in work teams. *Safety Science* 47(7). Elsevier: 1002–1006. http://doi.org/10.1016/j.ssci.2008.11.001

Chi S and Han S (2013) Analyses of systems theory for construction accident prevention with specific reference to OSHA accident reports. *International Journal of Project Management* 31(7). Elsevier: 1027–1041. http://doi.org/10.1016/j.ijproman.2012.12.004

Conchie SM and Burns C (2008) Trust and risk communication in high-risk organizations: A test of principles from social risk research. *Risk Analysis* 28(1). PubMed: 141–149. http://doi.org/10.1111/j.1539-6924.2008.01006.x

Conchie SM, Taylor PJ and Charlton A (2011) Trust and distrust in safety leadership: Mirror reflections? *Safety Science* 49(8–9). Elsevier: 1208–1214. http://doi.org/10.1016/j.ssci.2011.04.002

Erdem F and Özen-Aytemur J (2014) Context-specific dimensions of trust in manager, subordinate and co-worker in organizations. *Journal of Arts and Humanities* 3(10). LAR Center Press: 28–40. http://doi.org/10.18533/journal.v3i10.585

Felletti S and Paglieri F (2019) Trust your peers! How trust among citizens can foster collective risk prevention. *International Journal of Disaster Risk Reduction* 36(February). Elsevier: 101082. http://doi.org/10.1016/j.ijdrr.2019.101082

Gümüştaş C and Küskü F (2021) Dynamics of organizational distrust: An exploratory study in workplace safety. *Safety Science* 134(September 2020). Elsevier: 105032. http://doi.org/10.1016/j.ssci.2020.105032

Hajmohammad S and Vachon S (2014) Safety culture: A catalyst for sustainable development. *Journal of Business Ethics* 123(2). Springer: 263–281. http://doi.org/10.1007/s10551-013-1813-0

Kath LM, Magley VJ and Marmet M (2010) The role of organizational trust in safety climate's influence on organizational outcomes. *Accident Analysis & Prevention* 42(5). Elsevier: 1488–1497. http://doi.org/10.1016/j.aap.2009.11.010

Kożuch B and Sienkiewicz-Małyjurek K (2022) Building collaborative trust in public safety networks. *Safety Science* 152(April). Elsevier: 105785. http://doi.org/10.1016/j.ssci.2022.105785

Mearns K, Whitaker SM and Flin R (2003) Safety climate, safety management practice and safety performance in offshore environments. *Safety Science* 41(8). Elsevier: 641–680. http://doi.org/10.1016/S0925-7535(02)00011-5

Mehta AM, Tam L, Greer DA, et al. (2020) Before crisis: How near-miss affects organizational trust and industry transference in emerging industries. *Public Relations Review* 46(2). Elsevier: 101886. http://doi.org/10.1016/j.pubrev.2020.101886

Schwatka NV and Rosecrance JC (2016) Safety climate and safety behaviors in the construction industry: The importance of co-workers commitment to safety. *Work* 54(2). PubMed: 401–413. http://doi.org/10.3233/WOR-162341

Skeepers NC and Mbohwa C (2015) A study on the leadership behaviour, safety leadership and safety performance in the construction industry in South Africa. *Procedia – Manufacturing* 4(Iess). Elsevier BV: 10–16. http://doi.org/10.1016/j.promfg.2015.11.008

Tan H and Lim A (2009) Trust in coworkers and trust in organizations. *Journal of Psychology: Interdisciplinary and Applied* 143(1). Taylor and Francis: 45–66. http://doi.org/10.3200/JRLP.143.1.45-66

Vinodkumar MN and Bhasi M (2010) Safety management practices and safety behaviour: Assessing the mediating role of safety knowledge and motivation. *Accident Analysis and Prevention* 42(6). Elsevier: 2082–2093. http://doi.org/10.1016/j.aap.2010.06.021

Wu X, Yuan H, Wang G, et al. (2019) Impacts of lean construction on safety systems: A system dynamics approach. *International Journal of Environmental Research and Public Health* 16(221). MDPI: 1–16. http://doi.org/10.3390/ijerph16020221

Yiu NSN, Chan DWM, Sze NN, et al. (2019) Implementation of safety management system for improving construction safety performance: A structural equation modelling approach. *Buildings* 9(89). MDPI: 1–18. http://doi.org/10.3390/buildings9040089

Yiu NSN, Sze NN and Chan DWM (2018) Implementation of safety management systems in Hong Kong construction industry – a safety practitioner's perspective. *Journal of Safety Research* 64. Elsevier: 1–9. http://doi.org/10.1016/j.jsr.2017.12.011

Zhou Z, Goh YM and Li Q (2015) Overview and analysis of safety management studies in the construction industry. *Safety Science* 72. Elsevier: 337–350. http://doi.org/10.1016/j.ssci.2014.10.006

Zin SM and Ismail F (2012) Employers' behavioural safety compliance factors toward occupational, safety and health improvement in the construction industry. *Procedia – Social and Behavioral Sciences* 36. Elsevier BV: 742–751. http://doi.org/10.1016/j.sbspro.2012.03.081

Zohar D (1980) Safety climate in industrial organizations: Theoretical and applied implications. *Journal of Applied Psychology* 65(1). American Psychological Association Inc.: 96–102. http://doi.org/10.1037/0021-9010.65.1.96

Zohar D (2000) A group-level model of safety climate: Testing the effect of group climate on microaccidents in manufacturing jobs. *Journal of Applied Psychology* 85(4). American Psychological Association Inc.: 587–596. http://doi.org/10.1037/0021-9010.85.4.587

Index

For Product Safety Concerns and Information please contact our EU
representative GPSR@taylorandfrancis.com
Taylor & Francis Verlag GmbH, Kaufingerstraße 24, 80331 München, Germany

www.ingramcontent.com/pod-product-compliance
Lightning Source LLC
Chambersburg PA
CBHW060321220326
41598CB00027B/4386